T0201061

CAPTURING GLACIERS

WEYERHAEUSER
ENVIRONMENTAL BOOKS
Paul S. Sutter / Editor

Weyerhaeuser Environmental Books explore human relationships
with natural environments in all their variety and complexity. They seek to
cast new light on the ways that natural systems affect human communities,
the ways that people affect the environments of which they are a part, and
the ways that different cultural conceptions of nature profoundly shape
our sense of the world around us. A complete list of the books in
the series appears at the end of this book.

CAPTURING
GLACIERS

A History of Repeat Photography and Global Warming

DANI INKPEN

University of Washington Press / Seattle

Capturing Glaciers is published with the assistance of a grant from the Weyerhaeuser Environmental Books Endowment, established by the Weyerhaeuser Company Foundation, members of the Weyerhaeuser family, and Janet and Jack Creighton.

Design by Mindy Basinger Hill / Composed in Garamond Premier Pro

UNIVERSITY OF WASHINGTON PRESS   *uwapress.uw.edu*

LIBRARY OF CONGRESS CATALOGING-IN-PUBLICATION DATA
Names: Inkpen, Dani, author.
Title: Capturing glaciers : a history of repeat photography
and global warming / Dani Inkpen.
Description: Seattle : University of Washington Press, [2023] | Series: Weyerhaeuser
environmental books | Includes bibliographical references and index.
Identifiers: LCCN 2023034967 | ISBN 9780295752013 (hardcover) |
ISBN 9780295752020 (paperback) | ISBN 9780295752037 (ebook)
Subjects: LCSH: Glaciology—Methodology. | Glaciology—History. | Photography
in environmental monitoring. | Repeat photography. | Global warming—
Research—History. | Climatic changes—Research—History.
Classification: LCC GB2401.7 .I65 2023 | DDC 551.31—dc23/eng/20231011
LC record available at https://lccn.loc.gov/2023034967

∞ This paper meets the requirements
of ANSI/NISO Z39.48-1992 (Permanence of Paper).

FOR DREW

and that sunny afternoon

on the rainbow-pebbled beach

at the back of Bow Lake

Photography is motionless and frozen,

it has the cryogenic power to preserve objects

through time without decay.

PETER WOLLEN / "Fire and Ice"

CONTENTS

FOREWORD / *The Iconography of Ice*

PAUL S. SUTTER

What does global warming look like? While our world offers a growing array of dramatic images—floods, wildfires, extreme weather—with which to illustrate the effects of the invisible buildup of atmospheric greenhouse gases, our culture has persistently resorted to the iconography of melting ice. Why? One reason is that the incremental melting of glaciers, sea ice, and polar ice caps is among the most dramatic examples of how warming average temperatures are changing earth systems and how the poles and mountain landscapes are experiencing those changes in an amplified manner. Melting ice sits symbolically atop a watershed, with cascading downstream effects from a warming planet. Another reason is the scientific impulse to seek out answers to how "nature" is reacting to human climate, forcing in the planet's least anthropogenic landscapes. These places of distant ice most resemble the laboratory conditions scientists seek in measuring the signal of human-induced climate change without the noise of other human activities. From ice as a natural laboratory, it's not a stretch to see melting ice as a kind of disappearing wilderness, an end to forms of nature that had long existed beyond the reach of human activities. We are now well trained in such wilderness laments, even if climate change and melting ice are relatively new components of that conversation. Melting ice compels us because it can appear to be an instrument of measurement itself. A glacier receding back up a montane valley seems akin to mercury rising in the column of a thermometer; it allows us to visualize a process that is otherwise difficult to comprehend. For all these reasons, melting ice has become compelling visual evidence for global warming.

No project better embodies this powerful iconography of ice than the "Extreme Ice Survey" (EIS), which gained fame with the award-winning 2012 documentary *Chasing Ice*. EIS had its origins when *National Geographic* commissioned photographer James Balog to illustrate its June 2007 cover story, "The Big Thaw," with photographs of receding glaciers from around the world. As Balog recalled, the assignment unexpectedly became a "scouting mission," for he realized that the planet's icy nether regions, which he found aesthetically captivating, were changing rapidly as the climate warmed and with few people on the ground to bear witness. So Balog created a long-term repeat photography project. He and his team positioned and anchored more than two dozen cameras in extreme environments around the world to take regular time-lapse photographs of the termini and calving faces of glaciers over a multiyear period, creating a massive visual archive of glacial retreat. Balog is an artist, not a scientist (although he does have a master's degree in geomorphology), so he partnered with some of the world's leading glaciologists, including Tad Pfeffer of the University of Colorado's Institute of Arctic and Alpine Research, to design a project that would gather meaningful scientific data. Balog's project was also propelled by his sense of the scientific limitations inherent in communicating the invisibilities and abstractions of global warming. As he noted in *Chasing Ice*, "The public doesn't want to hear about more statistical studies, more computer models, more projections. What they need is a believable, understandable piece of visual evidence, something that grabs them in the gut." By visually capturing the world's receding glaciers Balog came to appreciate the gravity and immediacy of the problem, and he hoped to make his photographic visualizations an engine for similar realizations among a skeptical public. Seeing, Balog hoped, would be believing.

As Dani Inkpen demonstrates, in *Capturing Glaciers*, a vital and timely history of repeat glacier photography, James Balog's use of photographs to document glacier dynamics is nothing new. Repeat photography has a long and complex history as a technique in North American glacier science, one only recently tethered to climate action. We need to understand that history, Inkpen demonstrates, if we are to interrogate the iconography of ice that is such a dominant part of climate science's recent visual vocabulary. More than that, we need to think deeply and critically about photographs and how they have functioned as scientific evidence.

Photography was present at the birth of North American glacier science, to the good fortune of later repeat photographers. Most prominent among a group that Inkpen calls "glacier naturalists" were the Vaux siblings from Pennsylvania—Mary, William, and George—who turned to the study of glaciers during early family trips to the Canadian Rockies in the late nineteenth century and kept up a research program there until World War I. In the process, they became the first to systematically study the movements of North American glaciers, using photography and other forms of measurement as a critical part of their data-gathering.

Several points are worth noting about the Vauxes and the glacier naturalism they embodied. First, while they assumed that they were witnessing and documenting a general pattern of glacier recession, they did not implicate human activity in that trend. Rather, in the wake of nineteenth-century discoveries that ice sheets had once covered large parts of the continent, they assumed they were witnessing the last remnants of ice age retreat. The Vauxes also combined their glacier naturalism with mountaineering, mixing science and recreation in ways that would continue to shape glacier science through the century that followed, and their study of glaciers occurred in lockstep with the Euro-American settlement and development of the North American West. In this context, as Inkpen notes, their photographs not only functioned as scientific evidence of glacial dynamics but also served as "tourist mementos, advertisements, cartographic supplements, and representations of mountain wilderness." The Vauxes' particular iconography of ice thus defined glacial landscapes as primeval: empty of Indigenous peoples and their history and open to the hardy adventuring of late-Victorian naturalists. Their photographs lived on, not merely as evidence, then, but also as ideological expressions of a particular scientific milieu.

After World War I, glacier naturalism gave way to a new regime, "geophysical glaciology," characterized by a more quantitative approach, a shift to Alaska as a new locus of fieldwork, a spate of new technologies, and novel research questions, many of them influenced by European glaciology. Under this new regime, photography made less sense as a form of data-gathering. Ironically, it was a prolific photographer named William O. Field who was a key figure in this transition. Field's repeat photography work contributed to the realization that Alaska's tidewater glaciers were erratic in their movement and not uniformly shrinking, raising new questions about the relationship between glacier dynamics and climate. Before and after pictures were no longer enough, for rather than merely documenting an assumed trend of glacier retreat, they revealed anomalies that required closer study of glacier physics. "The photographic record generated by Alaskan glacier naturalists," Inkpen notes, "produced puzzles that could not be solved by more photographs." Geophysical glaciologists thus turned their attention to what was happening within the ice. Where glacier naturalists had studied easy-to-visualize glacier termini, geophysical glaciologists worked atop the névés, the accumulation zones where snow becomes glacial ice, using a new array of sophisticated and probing instruments to "see" inside glaciers. In this new regime, photographs no longer met the evidentiary standards of glacier scientists. It was at this point, Inkpen argues, that glacier science headed in a direction that James Balog would later lament—toward models, statistics, and the stuff of expert science that the public would struggle to understand. Glacier science became esoteric.

The years after World War II saw the rise of the two most important North American studies in this new geophysical mode: Project Snow Cornice and the Juneau Icefield Research Project, both of which were substantially supported by the US military. Photography did not entirely disappear from these studies, but it became less a way of studying glaciers and more a method for documenting the scientists and their work. Then, in the late 1950s, glaciology shifted again, led by a group of northwestern glaciologists centered at the University of Washington. This group, critical of geophysical glaciology's tight focus on internal glacier dynamics, began doing regional glacier studies. These scientists—and particularly the autodidact Austin Post—brought photography back into the field by using aerial photographs to do comparative glacier study and to scale up the micrometeorological work of the geophysical glaciologists.

From this synthesis of the geophysical and the regional would come a third regime, "environmental glaciology," which treated glaciers as features in and resources within a larger global environment. Environmental glaciologists were newly aware of the role that humans played in shaping climate; their work commenced just as Charles Keeling was documenting rising CO_2 levels in the earth's atmosphere from readings taken at the Mauna Loa Observatory in Hawai'i. But their science was about pushing beyond the physics of ice to understanding glacier dynamics in a larger environmental context, and Post's aerial photographs focused as much on advancing glaciers as they did retreating ones. In short, while environmental glaciologists would work out the relationship between glacier dynamics and global warming, particularly through a focus on sea level rise, they did not do so through photography, which was far too blunt an instrument to document the complexities of the climate-glacier relationship. Photographs still did not count as robust scientific evidence of global warming.

It was only in the early twentieth century that glacier scientists such as Bruce Molnia turned to repeat photography as a tool for showing that the world's glaciers were, as a rule, disappearing. They did so not necessarily in search of proof that this was happening—many other forms of more rigorous evidence already made that reality clear—but in search of a tool for communicating its complex science to a public who was showing itself to be remarkably impervious to esoteric forms of scientific evidence. This is where James Balog enters the story, as do a variety of other photographers, journalists, and mountaineers eager to communicate the perils of a warming planet. As Inkpen argues, their engagement with the iconography of ice did not make them bad scientists, even if repeat glacier photographs could be problematic forms of scientific evidence. But it did make them political actors, and in becoming political they inherited a form—glacier repeat photography—lacking the full consciousness of its history that *Capturing Glaciers* brilliantly imparts to us.

What, in the end, is problematic about the iconography of ice and the practices of glacial repeat photography that fed its rise? One set of problems is evidentiary; before and after photographs of receding glaciers present a tidy moral narrative, but they tell us little about the complex physical and environmental processes that determine glacier behavior. They serve as compelling but oversimplified translations of the arcane yet robust evidence that is the real stuff of contemporary glacial science. As poignant and motivating as these images have been for many, their simplifications have also created powerful openings for climate skeptics. Another set of problems comes in the framing. By centering the drama of anthropogenic climate change on alpine glaciers and polar ice, glaciologists have, in Inkpen's words, "reduced the problem of global warming to one of distant, unpeopled wilderness, a misrepresentation that is both incorrect and harmful." The story continues to focus on what the human species is doing to "Nature" rather than on what specific groups of humans through history have done to reshape the Earth's climate, with complex moral consequences. In the end, as *Capturing Glaciers* masterfully illustrates, the iconography of ice keeps questions of climate justice out of the picture.

ACKNOWLEDGMENTS

Many people have shaped and made possible this book. Much of my early intellectual debts are to my advisors and peers at the Department for the History of Science at Harvard University. Naomi Oreskes set the bar high and deftly guided the project's early stages. Sophia Roosth provided razor-sharp insights and support beyond the call of duty. Joyce Chaplin helped me see that more needed saying about mountains as Indigenous lands. Thank you to my then–fellow grad students who gave me support and feedback: Carolin Roeder; Louis Gerdelan; Yvan Prkachin; Miriam Rich; Colleen Lanier; Hannah Conway; Eli Nelson; Wythe Marschall; Devon Kennedy; Ben Goossen; Aaron Van Neste; Anouska Bhattacharyya; Alyssa Metzger; Anya Yermakova; and Leah Aronowsky. With a single perceptive comment, Gili Vidan made me rethink the significance of the whole thing. More than once Kit Heintzman provided thought-provoking comments on chapters at varying stages of intelligibility. An offhand comment from Steven Shapin was the stone that started the avalanche.

At the University of Washington Press, Paul Sutter, Andrew Bezanskis, and Mike Baccam steered the manuscript through some rough seas between 2019 and 2022, and they did so with discernment and solicitude. Marcella Landri: thank you for your diligence and patience in what seemed (at least to me) an endless chase of loose ends. Thank you to two anonymous reviewers whose generous insights helped make a manuscript a book.

It hasn't quite been a traveling road show, but this material has made the rounds.

Thank you to audiences at the University of Minnesota, Harvard University, the University of King's College, Cambridge University, Saint Mary's University, Dalhousie University, and Brandon University for your thoughts and questions. To my colleagues at Cape Breton University and Mount Allison University, thanks must go to you for hallway insights and feedback on work in progress. Charlotte Clarke provided comments on the initial first draft and invaluable research into Indigenous place-names. Thank you, Kiera Galway, for an eleventh-hour "shut up and write session," and Robbie Moser, for granting the use of his office. Brittany Jones swooped in to help me tie up loose ends during the critical, burnt-out phase.

Glaciologists have given me their time, memories, thoughts, and photographs, and they have been at the fore of my mind while writing this book. Carl Benson, Erin Pettit, Ethan Welty, Andrew Fountain, Bruce Molnia, and W. Tad Pfeffer: thank you all for sharing your memories and your knowledge of the history of glacier study. Tad, thank you also for your good-natured patience with my image requests. Wendell Tangborn will not see this book, but while writing I thought often of an afternoon in his sunlit West Coast kitchen. I am grateful to Birdie Krimmel for being there and sharing her memories too, and to Erin Whorton for making that afternoon happen and for sharing her experiences on the South Cascade. C. Suzanne Brown, thank you for the excellent photograph of Austin Post and for trusting me with your memories of Bill Field.

I would have gotten nowhere without the assistance of erudite archivists. My path through this material was made possible by Susan Peschel at the American Geographical Society Library Archives, Fawn Carter at the Alaska and Polar Regions Archives, Peter Collopy at the Caltech Archives, Anne Miche de Malleray at the Royal Swedish Academy of Sciences, and Elizabeth Kundert-Cameron and Kayla Cazes at the Whyte Museum Archives. Thank you also to Billie and Art for giving me a glass greenhouse to sleep in and bikes to ride while I was in Fairbanks.

This is a book about scientists, but I still have artists to thank. I am honored that Eirik Johnson shared his beautiful work. I am grateful to Teo Allain Chambi for accepting my poorly written Spanish request and sharing his grandfather's photograph with me. Thank you, Ben Pease, for accurate and beautiful maps. Charley Young, I wanted your art for the cover. Your images of the mountains we both love have made me rethink the ways we experience ice.

In recent months I have been fortunate to work with folks at the Canadian

Mountain Assessment, who have helped me see more clearly how scholarship can be a practice for cultivating good relations. This revelation (to me) impressed itself onto the final stages of this book in imperfect ways that leave room for improvement. Thanks especially to Graham McDowell, Madison Stevens, Daniel Sims, Megan Dicker, Keara Long Lightning, Tim Patterson, PearlAnn Reichwein, Nicole Wilson, Dawn Saunders Dahl, Karine Gagné, and Gabriella Richardson. Gùdia-Mary Jane Johnson helped me see there will always be work to do and that is just how it is. I have not lived up to the wisdom of all these wonderful people, but I will keep trying.

Dan Hallet is one of a handful of people who read the entire manuscript and is also ultimately responsible for inspiring the project: no words are thanks enough for bringing me to the mountains. Rook the dog was a steadfast writing companion, as was baby Gus in the revision period. My most substantial debt has been to my husband, Drew. His influence is woven throughout this book, from the images to the citations to the less tangible but indispensable work of keeping me going when I was sick of glaciers and cameras and Vauxes. He pushed the sled, which gave me the fortitude to keep pulling.

NOTE ON PLACE-NAMES
AND ENDONYMS

Throughout this work I have endeavored to use specific Indigenous place-names. This is not a history about Indigenous geographical knowledge. Nevertheless, the stories recounted here take place in present-day Alaska and western Canada, places of ongoing colonial history, and the stories themselves are part of this history. Using Indigenous place-names where possible and referring to specific peoples using their endonyms reminds us that Indigenous geographies still cover what Cole Harris, in *A Bounded Land*, calls a "bounded" land only partially subsumed by settler-colonial ambitions. I have not italicized words in Indigenous languages so as not to exoticize them as foreign in places where they have been spoken since time immemorial. For places known by multiple names, such as Everest—known as Chomolungma in Tibetan and Sagarmatha in Nepalese—I have stuck to one for clarity. I am certain I have missed opportunities and for that I apologize to those who have helped me see scholarship as an opportunity to work toward good relations.

CAPTURING GLACIERS

INTRODUCTION / *Thinking Historically about Photos of Ice*

I visited an old friend recently. It had been years since seeing the Bow Glacier. Both of us had changed. I was last in her neighborhood on a winter day so bright and cold it transformed my breath into crystals that shivered and sparkled in the air. She was indisposed, hibernating beneath her billowy robes of winter snow. I had to content myself with a view of her front garden, soft and rounded, blue and white. In summer she presides over one of the most breathtaking scenes on the (for now) aptly named Icefields Parkway in the Canadian Rockies. Perfectly framed by peaks, the glacier perches above the indigo waters of Bow Lake, to which she is connected by thundering Bow Falls and a creek that winds its way through rainbow-pebbled flats. The whole scene can be taken in from the front porch of red-roofed Bow Lake Lodge, set on the lake's shore by packer and guide Jimmy Simpson. In 1898 he deemed this a good spot to "build a shack."

I met the Bow Glacier the summer I left home, one of those free-spirited summers that Hollywood films coat thickly with nostalgia. Freshly released from the corridors of teenagedom, I chose a seasonal job that could not possibly advance the career I was preparing for in college but that would give me plenty of time in the mountains: housekeeping at a historic alpine lodge. In my time off I often scrambled up chossy peaks where I met wobbling marmots and grizzlies lounging in full-blooming meadows. I drank from swift, icy streams and camped wherever suited me (because, like many seasonal workers, I believed that national park rules

FIGURE 1 Old friends: The Bow Glacier and the author, 2003.

didn't apply to me). The Bow Glacier looked on with dignified indifference. I stood on her surface, secure in mountaineering harness and crampons (though a couple foolish times not) and marveled at the white westing plains of the Wapta Icefield from which the Bow drains, dreaming of even grander vistas beyond. I knew in those moments I was one of thousands to behold that sight yet felt like the world had just taken form. My happiness was untouchable, not yet complicated by the conundrums of adulthood. I was immortal; death did not exist and time would never run out.

But time does run. And glaciers, compressions of time in frozen water, are excellent gauges of its passage.[1] Mountain glaciers like the Bow are disappearing at rapid—and accelerating—rates. When Jimmy Simpson pondered building his shack, the Bow cascaded down to a forest abutting the lake in three undulating lobes, with the topmost flaring like outstretched eagle wings. I studied its shape from a black-and-white photograph hanging in the lobby of the lodge (fig. 1). Crevasse-torn icefalls separated the lobes, giving the glacier an intimidating look. It was big. It was beautiful. But the Bow Glacier has since receded. When I arrived one hundred years later, only the topmost lobe remained; dark cliff bands, wetted by Bow Falls, stood where crevasses once churned. The eagle wings were gone, and the glacier's surface was noticeably lowered. Yet you could still see its toe from the lodge. Today it has retracted even further. Like a wounded spider, it now huddles on the lip of the cliff over which it draped in 2003, barely visible from Bow Lake Lodge.

For many people who are not climate scientists, drastic recession of mountain glaciers like the Bow is clear and persuasive evidence of global warming.[2] Since most folks have never been to a glacier, photographs are often how they learn of disappearing ice. This is achieved through what are called repeat photographs: juxtapositions of old photographs and recent re-creations taken from the same perspective at the same time of year (because glaciers fluctuate with the seasons). Curiosity about the historical photographs in repeat series, like the one hanging in Bow Lake Lodge's lobby, eventually pulled my carefree summer in the Rockies into the trajectory of a professional life. This book is the result: it is about the people who photographed glaciers repeatedly and systematically to produce knowledge about glaciers and a variety of other subjects such as ice ages, wilderness, the physics of ice, and global warming. Throughout the twentieth century those studying

FIGURE 2 The Bow Glacier, 1902. This photograph is similar to the one
that hung in Bow Lake Lodge in the early 2000s. Whyte Museum of the
Canadian Rockies, Vaux Family Fonds, v653/NA1111.

glaciers used photography to capture changes in glacier extent and distribution,
but they did so for different reasons and with different consequences. I trace
the evolving motivations behind the use of cameras to capture images of ice and
concomitantly changing ideas about what is (or is not) being captured. The book
title is thus a double entendre, referring to both the enduring allure of glaciers
as repeat photographic subjects that "capture" beholders and the variety of ways
people sought to capture glaciers with their cameras. I pay especial attention to the
perceived value of repeat photographs as a form of evidence. Doing so illuminates
some of the ways repeat photography has encapsulated and conveyed changing
ideas about what glaciers are and why they matter. Photographs of glaciers are
about more than just glaciers. They're also about nature, land, how we can know
about such things, and the value we ascribe to them. Grasping this allows us to

better appreciate repeat glacier photographs for what they can tell us about global warming, but also how they are conditioned by history and where they fall short. It helps us see them not as static representations of the present situation, but as still-evolving elements in a process much bigger and more complex than any photograph could possibly capture.

I take a photograph-centered approach, following the photographs to archival information about the practices behind their creation. The history of how repeat photography was used to study glaciers in North America is checkered and discontinuous. Its value as a form of evidence ebbed and flowed based on ideas about what glaciers were and what knowledge-makers wanted to know. This was more than just a scientific matter. While many of the actors who populate these pages were scientists, producing knowledge of glaciers required an extensive host of characters and institutions. And the meanings of repeat glacier photographs broke the bonds of scientific intention and interpretation, drawing from and circling back to potent cultural associations. We will see, then, that the value of a form of evidence is conditioned by nonscientific elements, including political and practical considerations. Evidence, like objectivity, has a history. And history continues to make itself felt in the present.

Iconographies of Ice

Even as I scrambled about the Bow in the early 2000s, repeat photographs of shrinking glaciers began appearing frequently in the media. They were in magazines, on websites, in books, and in reports. These photographs were taken by scientists, environmentalists, and professional photographers, all of whom offered the widespread, rapid disappearance of ice as visual evidence that the global climate was changing at an alarming and unnatural rate. Referred to as "witnesses," "beacons," and "canaries in the coal mine of global climate change," vanishing ice seemed to carry an unambiguous message: global warming is happening, and it is happening now.[3] Photographs of receding glaciers became one of the most well-recognized visualizations of human-caused climate change. They became icons of global warming.

An icon is an image that exists in an intimate representational relation with something else: it is a symbol or emblem; it stands for the thing it represents.

The term's origins lie in the Ancient Greek Εἴκον, denoting likeness or image and typically referring to depictions of the gods. Centuries of violent debate ensued when icons entered the lexicon of Christianity, a faith that forbids graven images of the divine. Were icons sacrilegious? Instances of divinity materialized? Or merely devotional metaphors? Whatever they were, their potency could not be disputed and people killed and died over them.[4] The modern secular meaning of icon has thankfully lost some of the high stakes that stalk religious connotations. Yet they can be powerful nonetheless because icons come to stand in for what they are associated with. Uncle Sam is an icon of American patriotism because, for anyone familiar with the finger-pointing, top-hatted fellow, he simply *means* American patriotism. The association is instantaneous and does not depend on representational accuracy. So, too, with images of receding glaciers and global warming. "When I say a glacier has melted X amount," observes geographer and glaciologist M Jackson, "I'm also saying climate change has altered something X amount."[5] Photographs of retreating glaciers now stand in for global warming.

Icons, though rhetorically powerful, can be treacherous. All representations reveal and veil; icons tend to do so in extreme measure. Repeat glacier photographs today, as scholars such as Jackson caution, tell only one limited story: that of glacier diminishment.[6] More specifically, repeat glacier photographs portray glaciers as wilderness landscapes and global warming as something happening in distant, pristine places. This situates them within an aesthetic tradition that has permeated Western culture since the Romantic movement tore through it like a hurricane in the eighteenth and nineteenth centuries. The idea of "wilderness" offers havens of remote, unpeopled landscapes, places other than us and beyond our control, an antidote to all that is modern and corrupt.[7] Wilderness is beautiful, it is unsullied, and, so the logic goes, it is worth saving. Scientists have had additional reasons for focusing on what they see as wilderness. The history of ecology reveals a preference for "pristine" environments: sites of nature's "primitive conditions." Ecologists have historically sought regularities in nature and regarded human activities as external "disturbing conditions" to those regularities.[8] This research preference for ostensibly unpeopled landscapes provided an additional argument in favor of wilderness protection: it served a scientific purpose.[9] Similar orientations appear to be at work in glacier study.[10]

Wilderness has proved to be a powerful rhetorical tool for environmentalists and scientists. But it comes at a price. One trouble with wilderness is precisely that it makes nature seem far from home. Iconographies of wilderness place global warming elsewhere, visually detaching environmental concern from everyday life.[11] By framing global warming as something that is happening to a pristine glacier "over there" and focusing our attention on a future when there will be no ice, repeat glacier photographs distract us from the fact that people are already living with the effects of global warming. What of the millions of people in Bangladesh being battered by increasingly frequent cyclones? Or the residents of the Pacific island of Kiribati, whose fields are already being inundated and salinized by rising ocean water? How to account for the folks in Colorado, whose homes were destroyed by a late December wildfire? Global warming is already impacting peoples' lives in diverse ways that are difficult to combat, yet somehow repeat glacier photographs point our attention elsewhere.

While it is not the responsibility of repeat glacier photographs to depict all aspects of global warming, by functioning as a dominant icon they overshadow other stories and lack important details. Repeat glacier photographs don't tell you what you can do about global warming. They don't tell you its causes or who is to blame. Rather, they partake in what has been diagnosed as a long lineage of "mass-media spectacles of crisis."[12] They portray glaciers as romantic soon-to-be fatalities of a looming global catastrophe. They paper over the social inequities baked into the making and perpetuation of global warming, which continue to shape how some folks, some critters, and some lands are impacted more acutely than others. The increase in concentrations of atmospheric greenhouse gases is not a legacy that all of humanity shares equally. Certain nations and certain demographics (the wealthy, industrialized, often colonizing ones) are more to blame than others. Yet they are unlikely to bear the brunt of the impacts, at least in the near term. Those less able to marshal capital will feel the bite of weirding weather more severely than those who can shelter behind wealth.[13] For Indigenous people, particularly, this imbalance feels like a continuation of the colonial violence they have lived with for hundreds of years.[14]

Which brings us to another trouble with wilderness. Potawatomi philosopher Kyle Powys Whyte writes that visions of Indigenous lands as historically and presently unoccupied places undercuts the conditions that could encourage the flourishing of Indigenous cultures.[15] In representations of global warming, wilderness does exactly this while also failing to acknowledge how global warming is altering lands that are home to some who, despite great resilience, stand to lose the most.[16] Repeat glacier photographs are guilty on this charge. And, as we shall see, in North America their production historically relied upon and contributed to the displacement of Indigenous people. Injury to insult.

Boiled down, the problem with repeat glacier photographs is that they oversimplify both glaciers and global warming. They reduce our understanding of glaciers to a narrative of recession—or "ruin," to borrow Jackson's term—and global warming to something that afflicts pristine landscapes, not people, animals, plants, microbes, homes, and infrastructures.[17] Repeat glacier photographs can "create misunderstandings," write historian Mark Carey and geologist Rodney Garrard, because they fail to "explore why and for whom glacier retreat is a problem."[18] Moreover, as Carey suggests, their revelatory format encourages a too-simple dichotomy of "believers" and "skeptics," and implies that global warming is solely a problem for the natural sciences, as opposed to one to which the social sciences and humanities must also contribute.[19]

Although it may seem I've just thrown the book at them, this doesn't mean repeat glacier photographs are *bad* or that they should be done away with. It means there are problems with them that need to be thoughtfully considered.

Scholars have responded to this situation in two ways. Some have added breadth and depth to our understanding of global warming by seeking other representations that better capture the lived experiences of violence and inequity in our warming world. Forest fires, shattered levies, climate refugees, and drought-sapped soils are but a few of the now-familiar visual representations of global warming.[20] Others have sought more nuanced and accurate understandings of glaciers that explore the varied connections between glaciers and those who live near them. Historians and anthropologists have studied how communities understand their glaciers and how glaciers become embedded in the cultural identities of people living at their feet.[21] This scholarly work has shown that, despite their connotations of removed purity, glaciers are deeply embedded in human lives and they have politics. The politics of ice, it turns out, are of sublime landscapes and resource extraction, of transnational capitalism and local livelihoods, of gendered authority in extreme environments, and of disparities between the Global North and Global South.[22]

Yet among these efforts to deepen our understanding of global warming and of glaciers, few scholars have thought historically about how glaciers are represented. That is, few have asked how representations like repeat glacier photographs have been made and understood over time.[23] Failing to do so means we miss an opportunity to wield a powerful tool. Historical inquiry can bring to light the specific people, events, contexts, and assumptions that have fed into the making of a current situation. Doing so provincializes the present and makes us rethink what we take for granted as having always been the case, perhaps opening a door to imagining how things might be otherwise.[24] Such a rethinking is what motivates this book. Rethinking the history behind the repeat glacier photographs that operate today as icons of global warming situates these images as products of contingent events and processes, giving us a better understanding of why they are the way they are and indicating how things could be different.

My story is a thus a friendly critique of the use (and sometimes abuse) of repeat glacier photographs. I argue that these icons carry their histories with them, but they do so in nontransparent ways that require investigation to uncover. Historicizing these visuals deepens our understanding of the critiques lodged against them. It also reveals resources for thoughtfully recognizing their limitations and how they have shaped our perceptions.

Historicizing Repeat Glacier Photographs

In North America, glacier photography is as old as scientific glacier study itself, and images of receding ice have played changing evidentiary roles throughout the history of glacier research. The photograph of the Bow in figure 2 predates scientific consensus on global warming by several decades, so the person who took it could not have intended to capture evidence of global warming. Fair enough. That is unsurprising. But even in the very recent past, glacier photographs have carried connotations that were other than environmental, to say the least. As late as 1962 Humble Oil (later ExxonMobil) used the potential to melt glaciers as a celebratory metric for the huge amount of energy the company produced. In one *Life* magazine advertisement from 1962, against a backdrop dominated by a brilliant blue tidewater glacier front, the company crowed that, "Each day Humble supplies enough energy to melt 7 million tons of glacier!"[25] It startles our contemporary sensibility to think of a time when fossil fuel companies could trumpet their ability to melt glaciers. These glimpses of other ways of seeing glaciers prompts the question: What did repeat glacier photographs say before they told of global warming?

This book does not aim to explain how repeat glacier photographs *became* icons. That is another story, a story about media and the way information circulates and is interpreted, and also a story of public perceptions of science. While my story will brush up against this one, mine is concerned with understanding how photography was used to know glaciers and what knowledge-makers believed about their objects of study. Photographs of glaciers meant a variety of things during the twentieth century. At times, they meant little at all. A widely used form of visual evidence at the turn of the century, photography was less frequently deployed in glacier study at midcentury, taking a back seat to other forms of evidence before being picked up again in the early 2000s. The arc of this story, then, is an offset spiral.

Understanding what occasioned the return—why photographs were accepted as evidence at one time, rejected at another, and taken up again more recently—requires attending to photography as both a tool of field science and a form of evidence, placing it among the changing techniques, methods, research agendas, and values of glacier study in the twentieth century. Doing so reveals how ideas about evidence and meaning are not pre-given but are situated and responsive to their time. It also reminds us that photographs contain their own limits. As art historian Kaja Silverman has observed, following cultural critic Walter Benjamin, the very frames of photographs indicate all that they cannot capture.[26] Photographs, even series of repeated ones, cannot disclose the stories of their making and meanings. That's why we need history.

First, a Bit about Glaciers

A history of repeat glacier photography requires an understanding of a few things about glaciers. In their basic outlines they are simple. A glacier is a large perennial body of snow, ice, and accompanying rock detritus that flows downward under the weight of its own mass.[27] They form when the accumulation of ice outweighs loss over a number of years. The part of the glacier where snow accumulates is called

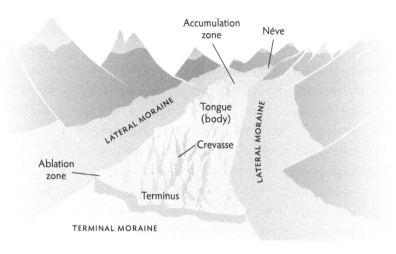

FIGURE 3 Anatomy of a mountain glacier. Drawing by S. Andrew Inkpen.

the "accumulation zone," or névé. At the névé, layers of precipitation collect and compress into a coarse-grained snow called "firn," which, under increasing pressure from new layers above, transforms to ice at greater depths. The deeper into a glacier you go, the older, harder, and darker the ice becomes, as air bubbles are squeezed out, creating larger crystals that absorb red light and reflect blue light. Midnight blue, even black ice can be thousands of years old. Movement is a defining trait of glaciers and is found in their very name: *glac*, from the Latin word for ice, and the suffix *ier*, which connotes conveying or transporting something.[28] A glacier transports ice down its body (or tongue) to a lower region, where ablation occurs during warmer months. Ablation simply means loss of ice. It can occur through melting, evaporation, and calving. Calving is the most dramatic way a glacier can lose ice; it happens when large pieces fall off the glacier's terminus, or end point.

A few words on glacier typology: A body of glacier ice is an **ice sheet** if it is continental in size, like that found in Antarctica. An **ice cap** is a smaller version of an ice sheet—how it flows is not directly influenced by underlying topography. An **ice field** is like an ice cap but smaller and its flow is influenced by topography. A **mountain glacier**—our subject matter—flows out of an ice field that spans several peaks. Mountain glaciers are further subdivided as cirque, hanging, rock (covered in rubble), and valley (which pour down in long tongues well below the summer snow line). A **tidewater glacier** is a valley glacier that terminates in salt water. A landscape that is presently covered in glaciers is **glacierized**, and a landscape that was previously shaped by glaciers is **glaciated**.

Glaciers are crucial links in the hydrological cycle that sustains life, human or otherwise. They accumulate and store precipitation, then release it during the late warm season when other sources of water dry up. It is thus worrisome when glaciers shrink. They shrink when climatic conditions change and provide inadequate precipitation, or not enough of the right kind of precipitation—snow—than is lost through ablation or when the factors contributing to ablation, such as ambient air temperature, relative humidity, or albedo (the capacity of a surface to reflect solar energy) amplify the rate at which ice is lost to a degree that outstrips accumulation. The majority of the world's glaciers are presently receding.[29] Not only are their termini retreating, but their surfaces are lowering. Yet glaciers are complex beasts—the response of any single ice stream to climatic change is mediated by the complexities of its location, source, terminus, topography, and internal dynamics. Large ice sheets take a long time to show a response to climate. Comparatively petite mountain glaciers react more quickly. This ability to swiftly register climatic change makes mountain glaciers good diagnostic tools for assessing global warming. Between 1995 and 2001, changes in mountain glaciers in North America's northwestern corner alone contributed more to sea level rise than did the continental ice sheets of Greenland and Antarctica.[30]

But that is not the only reason we ought to think about mountain glaciers. Although they only account for 10 percent of perennial ice, mountain glaciers are generally more accessible than polar ice sheets. Scattered across the world's uplands, they are intertwined with human ways of life in multiple, sometimes incongruent ways.[31] They provide fresh water and regulate its flow to downstream communities, releasing water during parched months when it is most needed. Their meltwater sustains in other ways too. For instance, by keeping stream temperatures cool for fishes. In the Pacific Northwest this cooling is crucial for salmon, a culturally, economically, and spiritually valued food for the Nez Perce Tribe, the Yakama Nation, the Confederated Tribes of Warm Springs Reservation, and the Confederated Tribes of Umatilla Indian Reservation. Returning to spawn in their hatchling pools, Pacific salmon bring energy gathered from bountiful oceans and stored in their muscled, writhing bodies into interior ecosystems, where it is snatched up by hungry bears, people, ravens, and microbes. When stream temperatures rise above that which salmon eggs can withstand, the future of this ecologically and culturally vital energy transfer is in question.[32]

People engage with mountain glaciers differently. For this reason, anthropologist Julie Cruikshank posits, they are "good to think with," for they disrupt bipolar conceptual fields (like nature-culture and science-stories). They prod us to think differently about the challenges we face with a warming, destabilizing world.[33] Some people, such as the Lingít Aaní (Tlingit) of the northern Pacific Coast, regard them as members of the social order: fellows in a society populated by humans and beasts, plants, geological features, and spirits. Others treat them as objects deserving of community care, as do the Ladakhi of the Karakorum.[34] Recreationalists the world over have come to love mountain glaciers through mountaineering, skiing, and other forms of alpine sport and tourism. As the draw for such activities, mountain glaciers bring revenue to local and national economies. They can be very good neighbors. They can also be hostile neighbors. Outburst floods, or *jökulhaups*, occur when ice-dammed meltwater bursts from its confines, sending tsunamis of ice, water, and rock plummeting onto homes and infrastructure. In southeastern Iceland, generations of such catastrophes destroying homes and farms have carved scars into the social fabric. For those whose forebears suffered loss of life and property at the foot of a glacier, the current recession of ice signals much welcome safety and security.[35] In other cases, warming means less stability. The Peruvian town of Huaraz, which in December 1941 was inundated by glacial Lake Palcacocha, will likely experience more frequent *jökulhaups* as glacierized landscapes become more volatile in a warming world. How such events will transpire will be shaped by social relations among people and among people and glaciers.[36] Glaciers are important, but there is no single relation between people and ice; not now, not in the past.

Glaciers, Photographs, and History

A visual history of North American glacier study uncovers new ground in the history of science. Much of the current historical work on glacier study has focused on Europe and on ice age theorists like Louis Agassiz, whose work helped widen imaginations to the possibility that humans could alter global climate.[37]

The scientists who studied the movements, dynamics, and nature of glacier ice are generally thin on the ground among histories of global warming. Were they latecomers to the scientific consensus, as some have suggested?[38] Or have we simply missed them? Tracking the practices, research agendas, and technologies used by North American glacier photographers helps to answer this question. It reveals that we missed them.

Photography is a good entry point into the history of glacier science. Although there are instances of painting being used to document and study glaciers, once photography was made quicker and more portable it became an important tool for field science.[39] This process began in the later decades of the nineteenth century with the invention of dry plate technology in the 1860s, and was taken further by the lightweight, flexible film of George Eastman's Kodak camera in 1888. The Kodak camera did away with heavy, brittle glass plates and messy on-site developing chemicals.[40] Photography soon became the preferred way for researchers to visually document glaciers. Indeed, scientific repeat photography began with photographs of glaciers. It was first used to document changes in Alpine glacial landscapes in the nineteenth century.[41] Thenceforth, photography and the study of glaciers developed coevally.

Throughout history, knowledge-makers have used visual representations in the production, evaluation, and dissemination of knowledge, and historians of science have had much to say on the topic. Scientific representations can help create communities of specialists who speak the same "visual language." Visual representations can be used to signal and cultivate scientific virtues like objectivity or they can work as conveyances for normalizing ideas about race, gender, and class.[42] Scientific visualizations can even generate ways of seeing nature. Nancy Stepan writes that visualizations "reflect political, aesthetic and other projects that have the capacity to expand, or limit, our imaginative engagement with the natural world."[43] Indeed, repeat glacier photographs have as much to say about imaginings of nature—what it is, who it is for—as they do about glaciers or global warming.[44] Thinking historically about how visualizations have encouraged certain ways of seeing glaciers and global warming is the central concern of this book. I track the acceptance, rejection, and then re-uptake of a single kind of visual evidence: the repeat glacier photograph. This story thus brings into focus the historicity and specificity of evidence: the ability of particular visuals to operate as evidence depends on whom scientists are talking to and what they are trying to achieve. These changes in repeat glacier photography's evidential status align with changing ideas about glaciers and nature because these ideas were, in part, a product of the assumptions and intentions of camera-wielding knowledge-makers. Scientific visuals operate beyond the pages of technical essays and the corridors of academe. Against this historical backdrop of diverse meanings and intentions, the virtues and drawbacks of repeat glacier photographs—what they show, what they don't show, and what they imply about glaciers, knowledge, nature, and global warming—emerge vividly.

Bringing a historical eye to the study of glaciers demands caution in the selection and use of terms. Every instance covered in the following pages falls under the umbrella of "glacier study," while "glaciology" is reserved for a historically unique form of glacier study that emerges in the late interwar period. So, too, does "glaciologist" imply a particular kind of practitioner—one who does glaciology; whereas "glacialist" refers to earlier students of glaciers. The characters who people these pages are not the only ones to have photographed mountain glaciers.[45] Yet they are some of the most prolific and systematic glacier photographers of the twentieth century (give or take a couple decades) and their careers closely track contemporary developments in the study of glaciers. They need not to have been "professional" scientists (here to mean those who are paid to do science). Both *professional* and *scientist* are slippery terms—their meanings shift and they don't always track knowledge-making practices of interest.[46] This is why many historians of science focus on knowledge-making practices directly, irrespective of the titles held by the persons performing them. To be considered in this book, then, a person need only to have systematically used photography to study North America's glaciers during the twentieth century. They do not have to meet some contemporary notion about who counts as a "real" scientist.

First spurred to curiosity by the photographs of the Bow Glacier hanging in Bow Lake Lodge, I have used historical photographs as trail markers in my journey to understand representations of glaciers. The path they indicated winds through three regimes of research: glacier naturalism; geophysical glaciology; and environmental glaciology.[47] Each of these regimes was a constellation of practices, tools, research agendas, social alliances, and geographical arrangements that, taken together, characterized different ways of seeing and studying ice. They emerged sequentially, but the newcomer did not neatly eclipse its predecessor. Their bounds roughly correspond with other historians' periodizations that trace a twentieth-century shift from natural historical to environmental ways of making knowledge, highlighting similar elements (empire, military influence, and global thinking) as important for carving up twentieth-century thinking about the Earth.[48] The field practices and standards of evidence of each regime were variously shaped by mountain recreation, prevailing nationalist ideologies, regional geographies, patronage, foreign influence, and disciplinary alignments. As scientific agendas and field practices evolved, the role of photography changed and the meaning and import of glacier photographs changed with it. The photographs, then, illuminate intertwined histories of meaning and evidence and of socially and culturally embedded scientific practices.

The photographs lead first to Canada's western provinces of Alberta and British Columbia, where the earliest systematic photographic studies of glaciers in North America were done. Here, in the late-nineteenth century, the Canadian Pacific Railway (CPR) brought well-heeled tourists and mountaineers to the feet of huge glaciers. Some of these tourists were professional naturalists and geologists; others were amateurs with a penchant for natural history and alpine exploration. They practiced what I call *glacier naturalism*, which wedded photography, mountaineering, and natural history. In glacier naturalism repeat photographs were evidence for the retreat of ice ages and contributed to colonial reimaginings of mountain lands.[49]

From the Canadian Rockies the photographs lead to southeastern Alaska, where huge tidewater glaciers posed tricky puzzles that prompted scientists to leave the terminus and begin studying the névé—the upslope area of accumulation. This change in terrain, in conjunction with the influence of European research agendas, prompted new questions. A new regime of glacier study—*geophysical glaciology*—developed to answer them. In geophysical glaciology, glaciologists used sophisticated technologies and techniques to investigate the physical processes by which glacier ice moved and transformed. This presented greater logistical and financial challenges than did glacier naturalism. So, like many American earth scientists in the late twentieth century, glaciologists turned to the US Office of Naval Research for patronage. In a confluence of interests generated by the Cold War, the military and scientists alike wanted to know more about the structures and dynamics of snow and ice. This was not without consequences for glacier photography, which became one of the many tools glaciologists employed in the field, though deemed not particularly helpful. Instead, glaciologists turned their cameras toward documenting work and life in the field for their military patrons, depicting glaciers as a domain for military-sponsored technical science.

Why, then, the return to repeat photography at the end of the century? Whence the recent iconography of ice? In what I term *environmental glaciology*, glaciologists participated in the turn toward environmental monitoring that swept the earth sciences in the 1960s and 1970s. They came to regard glaciers as hydrological resources that, just like forests, arable land, and mineral deposits, needed to be monitored. While we might anticipate that this would occasion the return to repeat photography, glaciologists remained loyal to the values and approaches of geophysical glaciology, which devalued photographic evidence. Repeat photography, although it enjoyed a small revival in aerial perspectives, was not the preferred means for monitoring glacier resources. It was not until the late twentieth century that those studying glaciers again embraced the power of repeat photography. Yet this return was not prompted by glaciologists' epistemic concerns. That is, they were not motivated by a desire to make better knowledge using the standards of their scientific discipline. Instead, they acted on a perceived need to cast off technical language and communicate beyond the silo of their expertise to reach nonscientific publics, among which denial of global warming was growing. In doing so scientists returned to a technique that they had largely abandoned in favor of what they considered more sophisticated evidence. Today's iconography of ice emerged in part from the realization that the "best" evidence, scientifically speaking, is not always the right evidence for the situation—in this case, a state of widespread doubt and denial about the reality and dangers of global warming.

While it may have seemed to be the right evidence in the face of climate denial, repeat glacier photography came with connotations that scientists did not foresee and conditioned how people would understand both glaciers and global warming. These ideas were grounded in what had come before. Both repeat photography itself and the form of the photographs were shaped by the prior history of glacier study, and not without consequence. Repeat glacier photographs begin with a foot in the past. It is unsurprising, then, that the past makes itself known through them, which is all the more reason to investigate histories of their making. The photographs of the Bow Glacier that I encountered at Bow Lake Lodge turned out to be a window onto much more than glacier retreat.

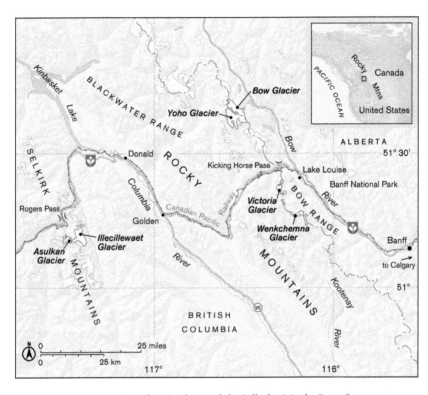

MAP 1 The Canadian Rockies and the Selkirks. Map by Pease Press.

1

DOCUMENTING / *Glacier Naturalism*

Banff, Alberta, celebrates its heritage in its street signs. Street names—Moose, Grizzly, Big Horn—remind visitors of the celebrity wildlife that drew tourists and big game hunters to Rocky Mountains National Park in its early years. On the corner of Bear and Buffalo sits the Whyte Museum of the Canadian Rockies, an earthy, midcentury modern structure of stone and wood that fits neatly into its surroundings. The Whyte is devoted to the mountain culture of the Canadian Rockies and houses (among other treasures) thousands of historical photographs. It is a good place to seek images of ice. One of the more extensive photographic collections is that of the Vauxes, a well-to-do Pennsylvanian Quaker family who frequented the area at the turn of the twentieth century and documented their travels in thousands of black-and-white photographs.[1] Siblings Mary, George Jr., and William Jr. first visited the Rockies in 1887 with their father as part of a family vacation to the West Coast. Along the way they stopped at Glacier House, a trackside hotel couched among the billowing peaks of the Selkirks, where they encountered glaciers of enormous magnitude. The Vauxes were among the first to systematically document the fluctuations of North American glaciers, and they initiated the first continuous study of Canadian ice streams.

The Canadian Rockies and Selkirks ("Canadian Rockies," for short) comprise the southeastern part of the Canadian Rocky Mountain Cordillera, an elevated, craggy region that stretches from the Pacific Coast in the west to prairie buttresses in the east and from the forty-ninth parallel in the south to the Alaskan border in the north (see map 1). They are part of the traditional lands of the Îyahê Nakoda

(Stoney Nakoda), Tsuut'ina, Kainai, Pikuni, Niitsitapiksi, and Métis Peoples.[2] The Stoney Nakoda, driven from the east by the horrors of smallpox, first came to what they called the Shining Mountains in the mid-eighteenth century. They made new homes in the eastern front ranges. According to Chief John Snow, the mountains were their sacred places. "In the olden days," as he explained in 1977, neighboring tribes referred to them as the "People of the Shining Mountains." "These mountains," he continued, "are our temples, our sanctuaries, and our resting places."[3] The Stoney Nakoda and their neighbors knew the mountains intimately and often served as guides to the voyageurs and settlers who began arriving. But they would eventually be blocked from pursuing their traditional activities in the mountains as the area was transformed into a park for tourists and naturalists like the Vauxes. The rise of glacier study in this part of Canada, then, coincides with the dispossession of Indigenous people. The colonizing and capitalizing forces that pushed out the Nakoda and others were the same conditions that made possible the study of the ice. The Vauxes' glacier photographs played a part of this upheaval.

To the west of the Shining Mountains, the Selkirks, the homelands of the Secwépemc (Shushwap), Ktunaxa (Kootenai Nation) and Syilx/Okanagan Peoples, were the abode of tremendous ice streams that wound down from rugged heights to nearly touch the railway's ties. It was an early site for mountaineering in North America.[4] The conjunction of ice and railway made this an excellent place for mountain climbing, which was then characterized by a preference for ice and snow rather than rock. The Canadian Rockies, with easily accessible glaciers, were dubbed the "Canadian Alps" and "Switzerland of America" for their resemblance to European ranges. A short hike from the imposing Illecillewaet (from the Okanagan for "big water") and Asulkan (Shushwap for "mountain goat" or "goat") Glaciers, Glacier House Hotel was an ideal headquarters for climbing and studying ice. A combination of transportation infrastructure and geographical happenstance brought altitude-seeking naturalists like the Vauxes to Canada's mountains, where they embarked on studies of the readily accessible ice streams. Over the years the Vauxes would return to Glacier House again and again, drawn by a love of alpine climbing and a curiosity about glaciers. The photographic legacy of their adventures eventually came to rest at the Whyte Museum and is now open to the perusal of passersby and researchers.

Copies of the Vaux photographs are tucked in clear plastic sheaths collated in

FIGURE 4 Toe of the Illecillewaet Glacier, British Columbia, Canada, 1902. Whyte Museum of the Canadian Rockies, Vaux family fonds, v653/NA-1354.

hefty albums that greet you when you enter the archives. Flipping through them, I came across a striking image: a head-on shot of the Illecillewaet, or "Great" Glacier, as they called it, taken in 1902 (fig. 4). In the photograph the glacier tumbles effusively downslope, funneled by steep lateral moraines on both sides and rupturing into cascading crevasses under the strain of its own huge mass. The pyramidal summit of Mount Sir Donald seen in the background is dwarfed by the lolling ice tongue, which rushes toward the frame as if it wished to peer over the photo's edge. Yet even in this grand state, the Illecillewaet showed signs of reces-sion. The height of the loose rock moraines bracketing it suggest the glacier was once larger. It was this combination of grandeur and diminishment that intrigued early glacier researchers like the Vauxes and brought them to the Canadian West.

Early glacier study relied on repeat photography to document glacier movements. There were also surveys and measurement techniques for comparing the behaviors of ice. The people who did this work were called "glacialists," and I use the term "glacier naturalism" to refer to the work they did. Glacialists were explorers, surveyors, geologists, naturalists, mountaineers, and tourists who came to the mountains for myriad, often overlapping reasons in the late-nineteenth and early twentieth centuries. Glacier naturalism was a branch of natural history, a mode of knowledge-making based in collecting, comparing, and cataloging specimens taken from the natural world. Natural history has historically had close ties with imperialism. Rooted in the Renaissance and Early Modern periods, natural history reached its apogee in the nineteenth century when European naturalists traveled the world "discovering," collecting, and ordering nature's diversity and dynamism, often with the guidance or appropriation of Indigenous knowledge.[5] Historians of science have paid close attention to the ways that Victorian naturalists used the social networks and transportation infrastructures of empire to facilitate their work. Charles Darwin, for instance, used his bourgeois social connections to land a position as captain's companion on the HMS *Beagle*, a British navy vessel that brought him to locales in the Southern Hemisphere, whence he gathered specimens that later helped him formulate his theory of evolution.[6] Glacialists were natural historians of ice but they did not collect samples of glaciers. Instead, they collected photographs. These served multiple purposes, operating as a scientific record of fluctuating glacier termini, tourist mementos, advertisements, cartographic supplements, and representations of mountain wilderness. Like other nineteenth-century naturalists, glacialists relied on the infrastructures of empire to conduct their research, and their work in turn shaped perceptions of the areas they investigated to align with imperial imaginings.

Glacier naturalism was a scientific and social practice that can be analyzed through the material, social, and political conditions that shaped the work of the Vaux family and their allies in the Alpine Club of Canada. The decades framing 1900 were an active time for glacier study. The fact that many of today's repeat glacier images recreate photographs taken at this time testifies to the bustle of the period. Glacialists documented and compiled records of glaciers and used this information to inform theories of climatic change. It was a heady time for such theories, following Swiss naturalist Louis Agassiz's energetic promotion of the ice age hypothesis and the ensuing debates in the second half of the nineteenth century. The core of Agassiz's "glacial theory" was the claim that the world's glaciers once were significantly larger. He was not the first to notice that glaciers likely were bigger in the past. Unsurprisingly, people living in the shadows of Alpine glaciers knew that their local ice streams had once been larger. But Agassiz was a big-time player in the world of learning. He had the scientific prestige and connections necessary to give the theory wide circulation. He began spotting evidence of previous glaciation everywhere and soon expanded the theory to encompass the British Isles, Ontario, and the northeastern United States. By the end of his life in 1873 he was convinced the whole world had once been enveloped in ice.[7] Photographs, like that of the Illecillewaet, served as scientific evidence for evaluating such theories. For glacialists working in the Rockies, photographs were also a currency that helped them secure funding and gain access to the mountain cryosphere and were representations that fed into a particular vision of the Canadian West as a place for tourists but not for other folks like Indigenous people or unskilled laborers. The Vauxes' repeat photographs, too, served as icons, not for global warming but for an idea of Canadian mountain wilderness. Whatever they may have intended, the legacy of the Vauxes' photographs was much more than just evidence of glacier retreat.

The Vauxes of Philadelphia

In histories of the Canadian Rockies, the Vaux siblings—Mary, George Jr., and William Jr. (fig. 5)—hold an important and celebrated place alongside Mary Schäffer, Arthur O. Wheeler, Byron Harmon, and other regional celebrities. The Vauxes were a well-positioned Pennsylvanian family with ample financial resources and leisure time. Descended from a prominent Quaker lineage, they could trace their ancestry to two brother-patriarchs who settled in the Philadelphia–Valley Forge region in the eighteenth century. Like others of their social standing, the family traveled extensively. Yet, spooked by events during an ocean voyage in his youth, their father banned sea travel for his children. As a result, the Vaux siblings spent more vacation time on the North American continent than many of their bourgeois peers for whom overseas travel would have been a rite of passage.[8] The children were socialized in the Society of Friends and received "guarded" educa-

FIGURE 5 Mary M. Vaux, William S. Vaux Jr., and
George Vaux Jr., circa 1887. Whyte Museum of the Canadian
Rockies, Vaux family fonds, v653/2 PA-1.

tions in the manner of traditional American Quakers.[9] Appreciation for natural
history was an important part of their upbringing.[10] They took up photography
as an art form that was particularly well-suited to natural history. Their glacier
photographs, then and now, function both as useful records of glacier fluctuations
and as aesthetic creations.[11]

The youngest, William (1872–1908), was the sibling most devoted to the study
of glaciers. Possessed of a technical mind, he graduated from Haverford College's
engineering program and spent his working years as an architect. The Engineers'
Club of Philadelphia provided a forum for many of his extracurricular presen-

tations and their subsequent publication, including his last and most polished piece, "Modern Glaciers." Although his older brother George appears as first author on their cowritten work, William did much of the background research and legwork. His notebooks are filled with detailed jottings and sketches and his correspondence includes letters to and from notable glacialists and geologists around the world.

Although he was the most avid glacier enthusiast of the three siblings, William was the least accomplished mountaineer. Despite membership in the Alpine Club of Canada (ACC) and the American Alpine Club (for which he served as treasurer until his death), William's time in the mountains was dominated by studies of ice. A lack of interest in first ascents did not, however, prevent him from acquiring elitist and proprietary attitudes toward the Rockies, the kind that has been associated with mountaineers since British climbers cultivated an elitism and disdain for tourists of the Alps during the "Golden Age" of mountaineering (1854–65). The Brits distinguished the mountaineer, who experienced mountain landscapes through active, laborious ascension, from "mere alpine walkers and sightseers."[12] William, in an 1898 article penned for the *Minneapolis Journal*, likewise haughtily distinguished between *his* relation to the mountains—that of the "real mountain lover"—and the attitude of most visitors who "generally stop over one night and walk up to the foot of the Great Glacier, stroll around the hotel, feeling that they have done all there is to do."[13] (Ironically, this accurately describes William's own first visit to Glacier House.) Like other members of the "Rockies Cult"—the name given to a set of wealthy Americans who frequented the Canadian Rockies at this time—William described the Canadian Rockies as his own special home away from home and felt that as a return visitor he had a privileged relationship to the place. Such feelings expressed a desire to identify as a particular kind of person—in this case, a mountain "local"—as much as they did a unique connection to a powerful place. In this William resembled the Golden Age mountaineers he echoed, who in fact differed little from the targets of their scorn. Mountaineers, historian Zac Robinson points out, are often just a specialized type of tourist.[14] Like Victorian mountaineers in the Alps, William's attitude is proprietary toward a land that was not his home.

When William died from tuberculosis at age thirty-seven, much of the glaciological work fell to George (1863–1927), who was almost ten years his senior and busy with a law practice in Philadelphia. George was also a Haverford alumnus, with an LLB from the University of Pennsylvania. The phrase "active citizen" describes him well. In addition to his law practice, he was appointed by Theodore Roosevelt to serve on the Board of Indian Commissioners and was involved in the Institute of Colored Youth and the Philadelphia House of Refuge. He was also a member of the Philadelphia Academy of Natural Sciences, the Photographic Society of Philadelphia, the Mineralogical Society, and, like his brother, the American and Canadian Alpine Clubs. George's first scientific passion was mineralogy. When the two brothers gave talks, George would typically cover general mountain geology and William would present on the strictly glaciological material. Later in life, when work and age tied him to the city, George became an armchair patron who financed far-flung geological expeditions. The mineral Vauxite, known only to exist in the Silo Veinte Mine in the Rafael Bustillo province of Bolivia, was named in his honor by one such expedition.

Mary (1860–1940) was the eldest of the Vaux children. When their mother died in 1880, Mary assumed the domestic responsibilities that defined her life until her marriage at fifty-four to the secretary of the Smithsonian Institution, Charles Doolittle Walcott, whom she met in the Canadian Rockies.[15] Like George, Mary served on the Board of Indian Commissioners. Ostensibly a watchdog institution meant to keep an eye on the Bureau of Indian Affairs, there is little evidence that it functioned as such during either of their tenures. Mary's biographer, Marjorie Jones, characterizes Mary's reports for the board as "disturbing and laced with disdain for Indigenous Americans."[16] Like many turn-of-the-century women of her race and station, she was both a victim and perpetuator of the prejudices of her time.

Of the three siblings, Mary was the most accomplished mountaineer, and her reputation was (and still is) celebrated by Rocky Mountain devotees. She was the first woman recorded as summitting any 3,000-meter Canadian Rockies peak, as well as the first white woman to explore the Yoho Valley and climb Abbot Pass, notorious as the site of the first climbing fatality in the region. Mary stood out. Her mountaineering costume was bold for its time and included knickerbockers and a buckskin jacket—a style liable to shock more genteel vacationers. One of her climbing companions reported that their party's arrival on a railway platform "seemed to afford as much astonishment and speculation as did the arrival of the

Tartarin in the lobby of the *Rigi-Kulm*. Miss Vaux, in buckskin and knickerbockers [. . .] was snap-shotted without mercy."[17] Perhaps Mary found the attention amusing; photographs often show her wearing a faint hint of bemused mirth.

In the mountains she felt in her element. She professed: "Of course golf is a fine game, but can it compare with a day on the trail, or a scramble over the glacier, or even with a quiet day in camp to get things in order for the morrow's conquest? Some how [*sic*] when once this wild spirit enters the blood, golf courses & hotel piazzas, be they ever so brilliant, have no charm, and I can hardly wait to be off again."[18]

Despite her climbing achievements, in many ways Mary's life was nevertheless structured by Victorian gender expectations. She was the only Vaux sibling who did not go to university. Instead, she was tutored in the use of watercolors, her talents earning her the sobriquet "the Artist" among fellow mountain devotees.[19] Although she had been taking photographs since the early 1880s, she could not join the photographic society to which her brothers belonged until 1895, when it opened to female members. This meant that in the early glacier studies Mary was typically responsible for the postproduction aspects of photography: developing and sorting. However, after William's passing, she assumed greater, and eventually sole responsibility for the glacier photography work.

Long after William died and George got tied down by professional responsibilities, Mary returned to the Canadian Rockies almost every year until her death in 1940. For her the mountains were a formative place where she could escape some of the responsibilities and restrictions of urban and societal life in Philadelphia. Writing to Walcott in 1912, two years before their marriage, she effused, "Sometimes I feel that I can hardly wait till the time comes to escape from city life, to the free air of the everlasting hills."[20] Like William, she expressed feelings of ownership over the mountains, confessing to Walcott, "Thee knows I feel a sense of ownership" over the Yoho Valley ("thee" was a standard designation in nineteenth-century Quaker parlance).[21] In the Rockies she enjoyed greater social and physical freedom than in the city; she found her life's work and her life partner there. Although her marriage to Walcott would eventually swing her into Washington's orbit, where she befriended First Lady Lou Henry Hoover, she remained a grand old dame of the mountains. This phase of her life was memorialized by her friend Mary Schäffer, a fellow Pennsylvania Quaker and explorer of mountains, who christened a peak after her: Mount Mary Vaux, whose long shale slopes drop precipitously into saxe blue Moraine Lake. But that was long after she and her brothers began the glacier studies.

The Glacier Studies

The Vauxes' first visit to the Illecillewaet resembled those of many other visitors during the first years of the Canadian Pacific Railway. In his diary entry for July 17, 1887, fifteen-year-old William made an unremarkable record of their visit to the Great Glacier. He noted they walked to the toe of the glacier, measured the temperature of a puddle on the ice, and snapped some photographs. It is hard to imagine a more prosaic report. There were no effusive descriptions of alpine scenery, no daring mountaineering feats.[22] At this point the Vauxes were mere tourists, of the variety William would later disparage. They had come to the Rockies via the only route that existed through the Canadian cordillera at that time, traveling from Vancouver to Banff on their way back east. Glacier House was a standard stop on this journey. They stayed close to the railroad but were able to visit the Illecillewaet because it was then a short jaunt from the hotel. Guests frequently made the excursion, braving the dense forest riddled with stinging devil's club shrubbery and demoralizing sliding alders. Stepping onto the Illecillewaet Glacier meant fording an icy rill that gamboled from its snout. Even conceding that William was not one for gushing descriptions, it would be wrong to characterize this initial visit as an instance of lightning bolt inspiration. In a letter to a friend, Mary recalled a tough hike on a poorly marked path, then waiting on the other side of the outlet creek while her brothers explored the ice.[23] (The mountain woman of buckskin and knickerbockers had not yet emerged.) The photos they took were tourist mementos that only acquired scientific significance later. One photograph would turn out to be of particular importance, that of Rock E, as it would later come to be known (fig. 6): a large rectangular boulder squatted in front of the ice.

The Vauxes began to distinguish themselves as repeat customers with a particular passion when they returned to Glacier House seven years later in 1894. William was then twenty-two and, like his siblings, had become an adept photographer and keen observer. And he wasn't the only one who had changed. The Illecillewaet had changed too (fig. 7). A "broad space of loose boulders" had appeared below

FIGURE 6 Edge of ice, Illecillewaet Glacier, July 16, 1887. Rock E is the large
boulder with George Vaux standing in front of it. Note the prow-like shape
of the terminus in the top right corner. Whyte Museum of the
Canadian Rockies, Vaux family fonds, V653/NG-956.

its terminus. In 1887 the tangles of alder grew within 20 feet of the ice; now there
was a much wider and bare stony gap between the ice and its terminal moraine.
They also noted alterations in the shape of the ice front, which had transformed
from "a great mass with steep sides" to a low-angled lobe that sloped "evenly till
it die[d] away altogether in the stream."[24] This shape, like that of ebbing waves,
signals withdrawal.

The visitors did not need to measure the glacier to know that its form and relation to the surrounding moraines had changed. The ice had retracted well beyond
the distinctive Rock E, leaving a field of boulders and scree in its wake. Almost
one hundred years later, using the Vauxes' material, glaciologists calculated that

FIGURE 7 Rock E, Illecillewaet Glacier, 1894. Taken with same lens and same position as 1887. Note the gently sloping ice in the top right corner, and the larger surrounding moraines. Whyte Museum of the Canadian Rockies, Vaux family fonds, V653/NG-377.

between 1887 and 1962 the ice had receded approximately 1,300 meters. When the Vauxes first beheld the Illecillewaet it was at its modern maximum.[25] When they returned seven years later it had already begun its retreat.

Such dramatic changes did not go unnoticed. The recession of the nearby glacier was a topic of fireside conversation at Glacier House. In 1888, a year after the Vauxes' initial visit, two British clergymen, Reverend William Spotswood Green and his cousin Reverend William Swanzy, came to Glacier House to climb and survey. The two men had support from the Royal Geographical Society and free rail passes from the Canadian Pacific Railway. They used this patronage to map the area and to ascend some of the Rockies' more impressive peaks.[26] The book that resulted from their exploits, Green's *Among the Selkirk Glaciers* (1890), was later a crucial source for the Vauxes. Yet the information about glaciers was buried deep within this colonial travelogue. Green's rambling prose described the train's route, the "dark green forests, rushing streams, purple peaks, silvery ice," the "Chinamen" working the railway, and the alpine flora encountered on his scrambles.[27] Many words were devoted to mosquitoes, alders, and the antics of Jeff the dog, an incorrigible hiking companion of Glacier House guests. Readers interested in glaciers must persevere to the end of this chatty narrative for two pages describing the reverends' work on the Illecillewaet. In travel narratives like Green's, glaciers often function as crowning aspects of alpine scenery or as terrain that sometimes enabled, other times hindered, mountain travel. For the climbing churchmen, glacier studies were part of a larger program of exploration and claiming mountain summits. This is characteristic of glacier naturalism, where physical conquest and rational inquiry, recreation and science, are part of the same endeavor.

Green and Swanzy's glacier studies, like early North American glacier naturalism generally, followed conventions for glacier work developed in Europe. Some of the earliest of this began, unsurprisingly, in Iceland, where glaciers and people were close and often uneasy neighbors. As far back as the 1600s, Icelandic naturalists were exploring their island's glaciers with an eye to cataloging and understanding them. Unfortunately, much of their work did not make it into broader European learned circles. The most extensive of these early contributions, Sveinn Pálsson's *Draft of a Physical, Geographical, and Historical Development of Icelandic Ice Mountain* (completed in 1795), languished unheralded in the offices of the Natural History Society of Denmark, to be published in piecemeal fashion until a full Icelandic translation in 1945.[28] Far from metropolitan centers of knowledge and subjected to various colonial regimes, the fate of these nascent efforts is a lesson in the importance of place, power, and connection in histories of science.

A better-known origin story for the European study of ice begins in Switzerland, where local farmers and savants documented the fluctuations of local glaciers during the Little Ice Age.[29] In the 1830s and 1840s, Louis Agassiz, James David Forbes, and others painted rocks and planted poles on Alpine glaciers to measure their forward motion. They measured glacier heights, depths, and widths, and performed surveys of their termini.[30] Their methods were brought to North American glaciers, and Green and Swanzy were early importers. In August 1888 they staked poles in the ice. These poles would move with the flowing ice and their positions could later be measured to determine the rate at which the glacier crept forward. Alas, when they returned twelve days later, the poles had toppled over because surface melt had expanded the holes. Undaunted, the reverends could still discern shallow divots where the poles had stood and used these to determine that the ice in the center had moved a whopping 20 feet, while that closer to the lateral moraines had crawled forward 7 feet. These were surprising results for a twelve-day interval. That was one speedy glacier. When the Vauxes made their own measurements of ice flow in 1894 it was in part to check the 1888 calculations.

Green and Swanzy also tarred a few boulders near the front of the ice with the hope that future travelers might use them to determine terminus movements.[31] Not long after, the Viennese geologist Albrecht Penck took photographs from a rock he marked "P5," which he later claimed corresponded to a rock the Vauxes used as a photographic station (Rock W). However, like Swanzy and Green's, Penck's observations also went sideways. Failing to bring something with which to accurately mark the positions of his observations, he had to resort to counting steps from the glacier's edge and taking compass measurements, not a terribly accurate way to measure distance.[32]

Given efforts such as these, a lack of reliable prior data points was a problem for the Vauxes when they arrived in 1894. In addition to the marks left by Green and Penck, there was a confusion of enigmatic hieroglyphics carved on the rocks in the moraine by hotel guests and trainmen. One of the more legible ones was marked on a boulder: "August 1890." A bit of sleuthing determined these markings had been made by British climber Harold Topham, who had claimed the first ascent of Mount Sir Donald (the large pyramidal peak behind Illecillewaet) and performed quick studies of the glacier while there.[33] Unfortunately, he did not publish his work, and when queried by William, reported that he had misplaced his notes.[34] Nevertheless, from Topham's boulder the Vauxes were able to calculate in 1898 that the glacier had receded 138 meters, an average annual recession of 17 meters per year since 1887.[35] A speedy glacier, indeed, but in the other direction.

Early efforts to study the Illecillewaet were inconstant, inscrutable, and difficult to reproduce. That is why the Vauxes are celebrated as the first to systematically study North American glaciers. However, when placed in the company of these fitful starts, we recognize their work as part of a wider—if poorly manifested—desire to document the motion and fluctuations of Rockies glaciers. Glaciers were charismatic, dynamic aspects of mountain landscapes that beckoned curious railway workers and travelers for whom natural history was part of the pleasure of mountain travel. Yet most people could not do justice to the changes in the ice. Either they were simply passing through and would not return or they lacked the means to make good observations. The Vauxes possessed the motivation, the necessary photographic skills, and the financial means and leisure time to return to Glacier House year after year, constantly improving their methods.

For eleven years they regularly returned to the Illecillewaet to observe, photograph, and survey changes in the glacier. They extended their activities to other ice streams along the CPR, and other glaciers such as the Bow and the Wenkchemna, which were further afield from the rail tracks. The Illecillewaet remained their best-documented case. They populated the boulder field at the glacier's toe with photographic and topographic survey stations with names like Rock E and Rock W. The photograph in figure 4 was taken from the latter. Today, it is known as Photographer's Rock, in their honor.[36] They mapped the glacier's toe and surrounding area, measured the speed of the ice's downward flow by planting flags in the manner of Green and Swanzy, and captured its yearly recession using repeat photography. Not every glacier received such thorough treatment. For other glaciers they mostly stuck to repeat photography to document retreating and advancing ice, hauling burly large-format field cameras across rocky moraines and charred valleys of brûlée.

The Vauxes made a point of framing their work as purely observational. A 1906 paper asserted they would make no attempt to *explain* the fluctuations of the glaciers they measured, but merely outline their observations.[37] Theirs were reports of fieldwork. This deliberate abstaining from theorizing was significant, given the state of the field. At the turn of the twentieth century, the topic of glacier motion was still smoldering from heated arguments among British naturalists in the 1840s about how glaciers moved. At the root was a disagreement between Irish physicist and science popularizer John Tyndall and Scottish physicist James David Forbes, both accomplished mountaineers and students of Swiss glaciers. They disagreed over how best to describe glacier motion, and their dispute drew opinions from many of the era's leading scientists. Forbes held that it moved like a semiviscous fluid, not unlike the slow creep of lava. Tyndall objected that Forbes's merely analogical reasoning fell short of the standards for natural philosophy and offered an alternative hypothesis: that glaciers moved down their beds as one gigantic mass, compelled by pressure and fused into a single object by a phenomenon known as regelation, whereby ice melts under pressure and refreezes when pressure is released. Much ink was spilled over their disagreement.[38] Alas, for naught, as there is no single accepted mechanism. Today glaciologists recognize many forms of glacier motion, including pseudo-plastic flow, basal sliding, glacier quakes, and internal deformation.

This dispute illustrates an important feature of glacier science: it is often about more than just the ice. Entangled with the question of whether a glacier moved like lava or not were ideas about method, disciplinary identity, institutional politics, and competing visions of the scientific self. Historians have demonstrated how Tyndall and Forbes each drew upon his mountaineering experiences and conceptions of heroic masculinity to support his claims about the nature of glacier motion. He who was willing to risk his life scaling precipitous heights was a more reliable and trustworthy knowledge-maker, or so John Tyndall, whose mountaineering skills surpassed those of his rival, would have it. In arguments over whose knowledge was best, detached objectivity was less important than manly performances of heroic physicality.[39] In a pattern that recurs throughout modern glacier science, the study of ice was more than just an excuse to go to the mountains, and mountaineering was more than just a means for getting to the glaciers. The two activities were linked through close cultural ties and the idea that, in extreme environments like high mountains, scientific authority can be augmented by physical prowess and gumption.

Of course, it wasn't all about who was the better climber. There were real scientific conundrums. Glacier movement may at first seem simple—glaciers generally move downward—but it quickly becomes complex. How a glacier moves is determined by the internal dynamics of the ice: its composition, the topography and slope of the bed, and the character of the terminus (whether tidewater or grounded on land, whether constrained by valley walls or spreading freely over flat ground). These complexities rendered the subject of glacier motion a contentious topic at the approach of the twentieth century, for which explanatory theories abounded.

Yet, the Vauxes professed lack of interest in the lot of them. When William presented "Modern Glaciers" to the Engineers Club of Philadelphia in 1907, he was asked for his thoughts on the physics (read: dynamics) of glaciers. He responded that he had purposely avoided the theory of glacier motion because, according to him, "there are no less than nine different theories which have been advanced."[40] Physicists, he held, had developed complicated theories to explain simple facts, but William kept out of the discussion, believing instead (like Forbes) that it was enough to think of glaciers as slow frozen rivers. What William didn't realize that this itself was a model with theoretical commitments.

The Vauxes frequently invoked frozen rivers to describe glaciers. But rather than worry about explaining *how* glaciers flowed like rivers, they chose to focus on providing more, and more reliable, observations from the field. The topic of glacier motion, they felt, was theory-rich but observation-poor: more would be gained by improving how glacier movement was observed. Their efforts were fruitful. Take, for instance, the technique they used to measure the rate of a glacier's flow. Rather than fashion markers out of wooden stakes (as Green had done), they commissioned the making of special metal plates to serve as markers that could be inserted into the ice. At the center of each plate was a hollow three-inch pipe—originally designed as a screw-in point for small flags, they found this feature more useful when the plates were turned upside down, allowing it to serve as an anchor in the ice. In July 1899, with the help of C. E. Cartwright, assistant to Canadian Pacific Division Superintendent and mountain guide Edouard Feuz, William and George placed nine plates on the surface of the Illecillewaet. Using rocks scattered about the surrounding moraines as transit stations, they triangulated the bearings to get a baseline on the ice that ran perpendicular to the downward flow of the glacier. They had hoped to use a chain to measure the line directly across the glacier, but the ice was so crevassed and crinnilated they had to resort to triangulation. Feuz and William then descended onto the ice to lay the plates along this line.[41]

On August 11 the original team (plus one purportedly useless and unnamed assistant) returned to determine the new positions of the plates. They set up transits on either end of their baseline and, with William and Feuz again on the ice, recorded the new bearings of the eight of nine plates that remained. With these measurements they then calculated how far each plate had traveled with the moving ice surface over the intervening weeks. They also measured vertical distances traveled, but this assumed a constant topography beneath the glacier—a highly dubious assumption, William pointed out. To improve this data point, they set another plate—number 9—on the tip of the glacier's tongue and used a nearby crevasse to access a vertical cross section from which to calculate the vertical motion, that is, until the piece of ice it sat upon detached from the glacier and ceased to move with the main body of ice. The Vauxes left for home in mid-August, but Cartwright returned on September 5 to again measure the positions of the plates. From his measurements and those made in August, the Vauxes calculated an average daily motion of the terminal ice that ranged from 2.56 inches per day for the plate closest to the glacier's northern edge to 6.79 inches per day for the plate nearest its center. This, they noted, was considerably less than—and presumably more reliable than—the distances determined by Green in 1888.[42]

This was as technical as the Vauxes' fieldwork got. Most of their time in the field was devoted to measuring termini and documenting positions with their cameras. Repeat photographs were (in their view) the most reliable way of documenting change in glaciers over time. For those of us familiar with snapping quick shots on a phone camera, this might seem like a simple matter, but it was hardly that. The Vauxes' Kodak camera and Bausch and Lomb lens, although more convenient than the wet plate setups used by earlier naturalists, still required tripods and a good deal of time and patience. Moreover, steep moraines and slopes of talus and scree flanked by thickets of alder and devil's club did not make for easy travel with bulky equipment. Certainly, the path to the Illecillewaet became increasingly worn during the years that they frequented Glacier House, but other glaciers were not so conveniently reached. Plus there was the unpredictability of the ice itself. As seen in their efforts to measure the vertical motion of plate 9, the very dynamism they sought to capture could impede their research. Rocks used as triangulation stations shifted or tumbled down moraines, avalanche debris buried former observation stations, and mountain weather, rarely true to a forecast in the days before satellite meteorology, was ever ready to wreak havoc on their plans.

In their glacier research, then, the Vauxes worked to render a chaotic and changing landscape into stable, mobile representations that could be compared over time. Their intention was to create records for future comparison. Yet, they were far from being masters of their environment, as the celebratory rhetoric of the time or hagiographic accounts might have it. Rather, they were tiny specks on indifferent and potentially dangerous terrain. The success of their research depended upon an array of individuals and institutions. Getting to the ice, collecting the photographs, and getting their photographs into circulation for people to see required a large support network of people and institutions. This network operated on regional, national, and international levels and the relationships of patronage and support that comprised it brought the Vauxes' work into collusion with the aims of others, notably a railway looking for advertisements and a nation seeking transcontinental legitimacy. The Vauxes' photographs came to serve as much more than just evidence of landscape change.

Glacier Naturalism, Glacier Nationalism

Glacier naturalism in Canada was made possible by the westward expansion of the Canadian Pacific, the nation's first transcontinental rail line. In 1885 the CPR managed to carve and blast its way through the Rocky Mountains while operating on a deficit budget. Part of its strategy for recuperating losses was to bring monied tourists to the stunning Mountain West. William Cornelius Van Horne, the railway's director, famously, if apocryphally, declared: "If we can't export the scenery, then we'll import the tourists." The company set up fashionably rustic hotels like Glacier House, which, nestled in the steep and snowy peaks of the Selkirk Range, attracted mountaineers from around the world. To fill their hotels and train cars the company embarked on an advertising campaign that reached east and south to Europe and America. It portrayed Canada as a frontier nation with unlimited sporting potential. And the well-heeled tourists who came brought with them Victorian sensibilities about rational recreation, believing that leisure pursuits ought to contribute to what they assumed was the inevitable improvement of "mankind."[43] The railway's luxury hotels were well-positioned to make the nearby glaciers convenient sites upon which to exercise such sensibilities. Glacier House, as George and William noted, "seem[ed] to form a natural station for [glacier] observation."[44] Turns out, it was also an ideal place from which to construct the idea of a nation.

The Canadian Pacific Railway occupies a singular place in Canada's national mythos. As a nation lacking a founder's day or a revolution's conclusion to mark its birth, Canada has tended to seek its national identity in symbols: hockey, the

"North," and the canoe.[45] The railway has been one of these symbols since its completion, when Canadian Pacific promoters claimed credit for the consummation of confederation. The idea, as expressed by an early railway publicist, was that the railroad "magically transformed a widely scattered Dominion into a prosperous and progressive nation."[46] Through its advertising campaigns aimed at potential settlers and tourists, the CPR aligned itself with images now seen as iconically Canadian: the fertile prairies ("the last best West" and "breadbasket of the empire"); the wild northwest; and the glaciated peaks of the Rockies. Subsequent histories, such as Pierre Burton's enormously influential *The Great Railway, 1871–1881*, elevated the railway—with its beneficent industrialists and heroic politicians—to the status of legend.[47] The railway was both cause and symbol of a nation unified across geographic diversity, having saved a fledgling nation from the clutches of an expansionist United States and zippering it together along a spine of steel rails.

Like many transcontinental railways, the story of the Canadian Pacific is one of industry bedding with government.[48] The railway conglomerate funded the reelection of John A. Macdonald's federal Conservative Party in 1878 after he was disposed by voters disgusted by a prior scandal involving government support for a foreign railway consortium. CPR contributions to the Conservative Party in the 1880s amounted to between $25 and $43 million in 2021 terms.[49] In return, the Macdonald government provided the Canadian Pacific with public funds and lands on either side of the tracks, which the company could sell for profit. The railway received the 2021 equivalent of $42 million, and approximately ten million hectares of land between Ontario and the Rockies, acquired from Indigenous and Métis peoples during the 1870s and 1880s through iniquitous treaties and the duplicitous scrip system.[50] Once the railway was complete, Macdonald assiduously aligned his government with the success of transcontinental transportation. Even today some Canadians still consider "Sir John A." the most pivotal of the "Fathers of Confederation," a leader among the "parade of solemn and slightly dyspeptic Scots" typically credited with the nation's founding.[51]

In the 1990s Canadian historians began to critique what Daniel Francis termed histories of "technological nationalism"—the idea that a transportation technology could build a nation. The Canadian Pacific, wrote Francis, with no lack of ire, was "built chiefly on the backs of Chinese coolie labour, using land obtained for almost nothing from the Indians [*sic*] and capital raised for the most part in

Britain."[52] Irreverent histories like Francis's, concerned with the social and cultural construction of nations, argued that the railway cannot be said to have united much of anything. In some cases it exacerbated and even inculcated divisions between east and west, urban and rural, Anglophone and Francophone.[53] Still, a myth with the railroad at its center, though untrue, proved compelling in certain strains of Canadiana, its place cemented in popular culture through the music of Gordon Lightfoot, Stompin' Tom Connor, and Gord Downey.[54]

Mountains were at the core of the Canadian Pacific's vision of Canada. The railway was officially completed on November 7, 1885, when company president Donald Smith (another in the parade of dyspeptic Scots, and the eponym for Mount Sir Donald) drove home the last spike at a ceremony at Craigellachie, deep in the mountains of British Columbia. Within months the federal government established Rocky Mountains National Park. It encompassed a small area around hot springs that had been "discovered" by Canadian Pacific employees near the Banff townsite. The springs had long been known to Indigenous peoples—the Stoney Nakoda used them as places of healing and of cleansing.[55] Mineral baths were all the rage in the late Victorian period and federal bureaucrats and railway executives alike saw a potential cash cow. The railway became the primary and most vociferous promoter of Canada's first national park.[56] By spring of 1886 trains were conveying tourists through Canada's new mountain park. The adamantine bond between railway and park did not falter until the 1920s, when a coach road between Calgary and Banff created opportunities for other forms of tourism. Yet, during the years of the Vauxes' visits, the mountains were understood to be the "Canadian Pacific Rockies," with the government-backed railroad playing a defining role in the park's image, landscape, and demographics.

The Canadian Pacific was invested in creating a place that could generate tourism revenue. It wanted to develop a park not for the purposes of wilderness conservation, but to centralize control of the lands around Banff in the hands of the railroad, restricting access to the mountains only to paying tourists and the railway-sponsored businesses that would cater to them. The government obliged. The first Dominion Parks commissioner, J. B. Harkin, noted that "tourist traffic is one of the largest and most satisfactory means of revenue a nation can have ... the commercial potentialities [of Banff] are almost startling."[57] The railway and the Dominion government shaped Rocky Mountains Park into an exclusive alpine playground for sportsmen and lovers of wilderness. In a pattern enacted throughout histories of national parks the world over, the correlate of this was that the park barred certain people.[58] Unskilled laborers and Indigenous people who had hitherto worked, hunted, gathered, and traveled in the Rockies were singled out for exclusion, allowed in only under strict conditions, either as curiosities during Banff Indian Days or when work needed doing cheaply.[59] Indigenous guides became less common after the railway began importing Swiss mountain guides in 1899.[60]

To cultivate the Rockies as a place for monied tourists, Van Horne enlisted photographers and painters to create promotional material for which they received free passage or reduced fares. Before the line was even complete, photographers William Notman, Oliver Buell, and Alexander Henderson were commissioned to document the railway's heroic construction and the spectacular mountain setting. Canadian Pacific patronage especially benefited painters associated with the new Royal Canadian Academy of Arts in Ottawa.[61] While the railway's patronage of artists is well-documented, its patronage of scientists less so. Yet science was no less important than art in the company's efforts to both establish Rocky Mountains National Park and turn a profit, particularly in the case of photography, where art and science could overlap. The railway aspired to the creation of a Canadian cordillera marked by routes, lines, and summit objectives. The mapping and documentary activities of glacialists like the Vauxes helped build that landscape. Glacier naturalism, with its reliance on photography and focus on topographic features of interest to mountaineers, was well suited to contributing to this goal. Its contributions to Van Horne's reimagining of the Canadian Rockies became iconic representations of the young park: a haven of wilderness for adventurous tourists.

Glaciers played an integral role in the project of attracting mountaineers because the Victorian mountaineer, outfitted with the unwieldly alpenstock (a long staff with an iron spike tip) and hobnailed boots, favored snow and ice routes. (Climbing rock would not be in vogue until the Edwardian period, and then only in select places.)[62] They considered the glacier-clad mountains of the Alps the gold standard for mountaineering and judged other ranges accordingly.[63] Yet, by the end of the nineteenth century, most of the major peaks in the Alps had been summited and European climbers were looking for new challenges. Some tried more difficult routes up already climbed peaks, while others headed overseas. The Canadian Rockies, more so than the American Rockies to the south,

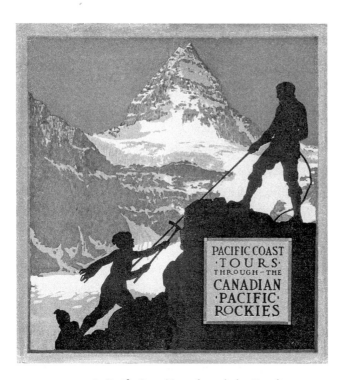

FIGURE 8 Pacific Coast Tours through the Canadian
Pacific Rockies, circa 1920. Note how the advertisement likens
the Rockies with the Alps through a likening of Mount Assiniboine
to the Matterhorn. CRHA/Exporail, Canadian Pacific
Railway Company Fonds.

were heavily glacierized, and thus offered just the right mix of the familiar and
the novel to European climbers. Although a far cry from Alpine chalets set amid
idyllic pastures, the railway's luxury hotels provided an appealing combination
of wilderness and comfort. American mountaineers, for whom a trip to Europe
was a rite of passage, also turned to the Canadian mountains for an almost-Al-
pine experience. The Rockies, crowed Canadian Pacific advertisements, were "50
Switzerlands in one!" (fig. 8).[64]

Visuals highlighting the Rockies' Alpine-esque glaciation aligned perfectly with
the Canadian Pacific's desire to market the park as a mountaineering playground
for wealthy urbanites.[65] The Vauxes' work supported this end. In 1900 they were

contracted to produce a promotional pamphlet devoted to the glaciers of the Selkirks and eastern Rockies. *The Glaciers of the Canadian Rockies and Selkirks* went through eleven editions. It was first William's responsibility, then George's after William's passing, and finally Mary's. In the 1911 edition Mary pointed to the mountains' likeness to the Alps: "Formerly," she wrote, "it was thought that if one would enjoy really fine mountains he [*sic*] must travel to Switzerland."[66] No longer! This informative and visually enticing pamphlet countered such misconceptions. Written as accessible natural history and generously illustrated with photographs, it asked, "What is a glacier? What are its distinguishing features? How does it act?" Answers, the authors informed readers, may be found by taking the Canadian Pacific Railway from Montreal to Vancouver. The pamphlet included illustrations of terminal moraines, the Illecillewaet and Asulkan Glaciers, a glossary of technical glacial terms, and descriptions of glaciers at Lake Louise, Field, and Glacier House—sites of the three crown jewels in the Canadian Pacific's chain of mountain hotels.

The Vauxes, like the artists in Van Horne's "Railway School of Art," Green and Swanzy, and the famed British mountaineer Edward Whymper, received complementary train passes with the understanding that they would share their creative products to attract "the type of tourist the Passenger Traffic Department coveted."[67] The railway was discerning in its choice of promoters. Free passage was given only to those recognized "for their celebrity or for their demonstrated ability" and whose efforts would reach the desired audiences. The Vauxes' easily fit the bill. Their photographs were included in shows of the Photographic Society of Philadelphia and the photo-succession movement's "Little Galleries" in New York. Their glacier papers were printed in national journals and presented before audiences of naturalists and engineers in Philadelphia. They cast a wide net over monied audiences in eastern America who, as Mary Schäffer, Walter Wilcox, Charles Fay, Charles Walcott, and the Vauxes themselves demonstrated, tended to repeatedly return to the Canadian Rockies. Between 1887 and 1898 nearly half the twelve thousand guests who signed the Glacier House guest book were Americans.[68]

The company gave the Vauxes more than just free passage. Railway employees of every rank gave time, labor, material, and information, both on and off the glaciers. Without the assistance of CPR mountain guides—such as Edouard Feuz's in 1899—much of the Vauxes' work would not have been done.[69] Employees like Division Superintendent C. E. Cartwright made measurements after the Vauxes had left for the season, while still others lent instruments, corrected transit measurements, and provided weather data. The Vauxes' glacier work, then, was supported in a number of ways by various units within the railway. For this they thanked their allies lavishly, inviting their benefactors and hosts to their exhibitions and furnishing them with gifts. Often these were photographic prints, but they could also be less conventional items such as the brass flagstaff they gave to the conductor of engine 738. Letters thanking them for their "beautiful prints" came from Molly Mollison, manager of Mount Stephen House, their guide, Ed Feuz, and from railway top brass.[70]

Through relationships of patronage and exchange with the Canadian Pacific Railway Company, the Vauxes' glacier photography was woven into the twin projects of profit-making and ideology-making: profit for the railway and an ideology of Canada based on the railway and mountains. Their work was further integrated into national mythmaking through their involvement with the Alpine Club of Canada and its fiery cofounder, Arthur O. Wheeler, who continued and expanded their glacier studies. Glacier naturalism thus became further aligned with claiming Canada's mountains as an exclusive space for rational recreation, frequented by wealthy tourists and facilitated solely by the railway.

A Nation of Mountaineers

The Vauxes' reliance on mountain guide Feuz illustrates a pragmatic connection: those who would study mountain glaciers needed mountaineering skills. Yet beyond the need of access to the field, mountaineering and glacier study were also connected in print culture: glacier naturalists used the writings of alpinists as source material for their studies and themselves contributed to that same literature. William Vaux's reading lists, one of which was generated by the famous climber Edward Whymper, included many mountaineering and travel narratives (including Green's) and articles published in journals such as *Appalachia*, the periodical of the Appalachian Mountain Club.[71] This is easy to understand. Mountaineers went places glacialists may not have been able to visit—through their wanderings and their writings they extended the geographical and temporal reach of glacier naturalism. From Charles Fay's *Appalachia* article on the first ascent of Mount Victoria, the Vauxes were able to determine the pre-1899 position of Victoria Glacier near Lake Louise. Walter Wilcox, who explored the heavily glaciarized Moraine Valley with his friend Samuel Allen in 1893 and 1894, sent William measurements and photographs and made suggestions for further study.[72] Glacialists encouraged these arrangements. Requests for observations from mountaineers appeared in the pages of alpine journals well into the twentieth century, as glacialists sought to enroll them as field assistants.[73] As late as 1949, even, some were calling the American Alpine Club an "auxiliary in Glacial Research."[74]

In Canada, the Alpine Club of Canada (ACC), founded in 1906, was the primary institution through which mountaineering served glacier study. Through it, both of these activities were woven into an ideology that promoted Canada as a "nation of mountaineers."[75] The club's origins betray its close alignment with the CPR and its nationalist aspirations. It was founded in Winnipeg, Manitoba, approximately 1,400 kilometers from the eastern edge of the Rockies. Hardly a mountain town (its apex reaches a paltry 240 meters above sea level), Winnipeg was chosen for its symbolic value. Lying roughly at the center of the country, the location could be reached by rail from both east and west. It was also the western headquarters of the Canadian Pacific.

The club's founders were journalist Elizabeth Parker and mountain surveyor Arthur Oliver Wheeler.[76] Parker had been recently enchanted by the mountains on a six-month visit to the Canadian Pacific Rockies. Wheeler knew the mountains through his phototopographical surveys, which involved taking photographs of a landscape from different vantage points and using the divergent perspectives to triangulate the distance between points.[77] Wheeler's summers were spent ascending peaks and ridges and fraternizing with the international set that frequented Glacier House—including the Vauxes' "Rocky Mountain Cult." It did not take long for him to become enamored with what he called "the delightful, devilish Selkirks" and to become painfully aware that Europeans and Americans were taking most of the first ascents.[78] His unlikely partnership with Parker was brought about by a shared desire to claim "Canada's mountains" for Canada. She too was concerned that foreigners were summiting Canadian peaks. In a racially charged appeal to the imperial sensibilities of a young nation, Parker chastised Canadian mountaineers: "It is simply amazing that we leave the hardships and the triumphs of first ascents to foreigners. Even a Hindoo Swami [*sic*] has climbed one of the highest peaks in this region."[79]

Historically, to make a first ascent was to lay a claim to the mountain. Often such claims were nationalist, as evidenced by nationally financed expeditions, by flag-planting summit ceremonies, and by the aftermath of the first ascent of Chomolungma (Everest), in which Britain, New Zealand, India, and Nepal all jockeyed to claim Tenzing Norgay's and Edmund Hillary's feat as their own.[80] Wheeler, an Irishman by birth, was a patriot of his adopted nation and saw mountaineering as an opportunity to express his particular form of patriotism, what historian PearlAnn Reichwein describes as an "Anglo-Canadian nationalism."[81] Wheeler's love of Canada was also a love of Anglo-imperial culture, and he channeled this love into promoting Canadian mountaineering on a British model. In 1906, with twenty free rail passes courtesy of the CPR president, a cadre of Canadian mountaineers, railroad officials, men of science, guides, and Parker and her daughter Jean, gathered at the Canadian Club in Winnipeg to found the Alpine Club of Canada.

Wheeler and Parker envisioned "a nation of mountaineers, loving its mountains with a patriot's passion."[82] They modeled their club on the oldest mountain recreation fraternity, *the* Alpine Club, established in London in 1857, mimicking its Victorian hierarchical classes and ladder of "graduating climbs." While European alpinists were exploring more challenging and technical routes and developing technologies that would allow them to climb steep rock faces, the ACC stuck to a style of mountaineering rooted in the tenets of Victorian practice: employ guides, stick to safe summit routes, maximize time on ice and snow, and restrict the use of technical aids (rope, nailed boots, and alpenstocks were fair game; pitons—metal wedges hammered into cracks in rocks to secure ropes—were not).[83] The club's social mores also tended toward conservatism. Women were admitted (unlike the London Alpine Club or the American club at the time), but Wheeler insisted they wear skirts at camp, changing into their knickerbockers only prior to setting off on a climb.[84] With a few notable exceptions, Jewish and Asian climbers were blackballed in their applications for membership. Likewise, Indigenous people and "ethnic" laborers had no place in Parker's and Wheeler's Rockies. Exceptions to this trend, ironically, only illustrated the club's imperialist leanings. When the talented Japanese climber Shozo Kitada was nominated for membership in 1929, Wheeler supported his application by claiming, "If anyone protests we should mention a Japanese Hon. Member of the English Alpine Club which is particular and generally considered pretty respectable."[85] In cultivating this bourgeois, patriotic-imperialist vision of the Dominion's mountains, the club reinforced the railway's idea of the park as a socially homogeneous place for the "right" kind of visitor.

Science was central to the ACC's vision of the Rockies. In her introduction to the first edition of the *Canadian Alpine Journal*, Parker stated the "*promotion of scientific study and the exploration of Canadian alpine and glacial regions*" was one of the club's main objectives.[86] She hoped that scientific explorers would populate the club's Scientific Section, particularly those interested in glaciers, alpine geography, and natural history. That was because knowing the mountains as objects of science was not just useful; it was a way of asserting a claim over alpine spaces. Wheeler confirmed this stance in his 1938 retrospective, "The Origin and Founding of the Alpine Club of Canada," according to which, the surveys paved the way through the mountains for the railway; the railway built Glacier House; Glacier House attracted the likes of the Vauxes and William Spotswood Green, who brought mountaineering and an enthusiasm for glacier studies; and this foreign dominance prompted a reply with the creation of the ACC. Mountaineering, glacier science, and the CPR were the three pillars of the early ACC.

Glacier study, and repeat photography especially, was a personal passion for Wheeler. He was introduced to it by the Vauxes at Glacier House in August 1902. "Met Mr. Vaux jr [likely George]," Wheeler noted in his diary, "and assisted him in taking measurement of Illecillewaet Glacier. Or rather, took measurements for him."[87] As a trained phototopographic surveyor, Wheeler was confident he could improve upon the Philadelphians' efforts. He began by taking repeat photographs and measurements of the Yoho Glacier near Mount Stephen House. When the club was established a few years later, glacier studies became part of the annual camps in which members gathered to climb and socialize. Studies were then published in the *Canadian Alpine Journal*. Couched amid narratives of mountain conquest and celebrations of Canada's alpine geography were scientific articles about Canadian glaciers—most of them written by Wheeler and the Vauxes.[88] In 1932 Wheeler built upon these efforts by organizing a section of the club devoted specifically to glacier studies. It was headed by himself but administrated through local chairmen who organized regional studies and collected data from members.[89] Wheeler compiled the data from the chairmen and individual glacier enthusiasts like New York member J. Monroe Thorington and published them in the journal. This framework ensured that the articles were funneled through Wheeler and conformed to his standards.[90]

By these means—the *Canadian Alpine Journal*, the annual mountaineering camps, and the Scientific and Glacier Sections—Wheeler and Parker enshrined glacier naturalism as an official element of the Alpine Club of Canada. Alongside mountain recreation, exploration, and art, it was a way of claiming Canada's mountains for Canada's alpine club. The ACC was not alone in its commitment to glacier study. The Swiss Alpine Club, the German-Austrian Alpine Club, the Italian Alpine Club, and the tourist clubs of Norway and of Sweden also incorporated glacier study into their clubs' activities. In Canada, the repeat photographs at the core of glacier naturalism did much more than simply serve as scientific representations, however. They were aspirational depictions of what Canada's mountains *should* look like at a moment when corporate and governmental forces were shaping them anew. Their success placed them within an iconography of Canadian mountain wilderness that persists today, one that presents the mountain parks as abodes of great glaciers and arenas for mountaineering and tourism.

Scaling Up

Glacier photographs taken by the Vauxes and other glacialists served the interests of a national railway hoping to attract mountaineers and those of an alpine club entangled with the railway and seeking to lay a patriotic claim on Canada's mountains. Their photographs circulated well beyond these national orbits and eventually were incorporated into the collecting efforts of the first international glacier monitoring organization: the Commission Internationale des Glaciers. Through the commission they became evidence for competing theories of global change and acquired a theoretical framing supported by the Vauxes.

The commission was the invention of an unlikely trio: François Alphonse Forel, professor of anatomy and physiology at University of Lausanne; Prince Roland Bonaparte, cousin of Napoleon III and amateur glacialist; and Marschall Hall of the Wiltshire militia and fellow of the Geographical Society of London. Formed in Zurich in August 1894 at a meeting of the International Geological Congress, the commission's stated purpose was to "encourage, and to collect observations on glaciers all over the world, with the special object in view of discovering a relation between the variations of glaciers and of meteorological phenomena."[91] Appointees from eleven member nations funneled glacier observations from individuals and organizations in their territories, where the commission compiled and redistributed them in the form of thirty-four annual reports between the year of its founding and 1927.[92] The quality of the data varied. The vast majority came from the Alps (particularly the western Alps), which became a standard against which to measure reports from other regions. The Scandinavian countries tended to submit extensive measurements made by scientists and polar explorers, while observations from mountaineers and surveyors in the Himalaya and Tian Shan were at best spotty and qualitative. Both Wheeler and the Vauxes contributed. In 1900 the entirety of one of the Vauxes' papers was included in the commission's report as an example of the high quality that glacier study could attain.

The commission was set up to ascertain the state of the world's glaciers: were they advancing or were they retreating? Were fluctuations the tail end of a great glacial epoch or caused by smaller cyclical fluctuations? At this time glacialists were enamored of the "glacier theory" promoted by Agassiz. The glacier theory provided a theoretical framework for interpreting photographs taken by the Vauxes and other glacialists. Against a prehistoric past enveloped in ice, "living" glaciers of the present seemed to be lingering fragments only. Like medieval Britons who marveled at ruined Roman architecture as *eald enta geweorc*—"the work of ancient giants"—glacialists saw in living glaciers the ruins of an epic geological past.[93] When the Vauxes looked at the Illecillewaet and Asulkan Glaciers they saw the vestiges of a much larger ice stream that had once encompassed both. Their observations and measurements, then, aimed to track the rate and nature of this diminution and bifurcation. Although uncomfortable with physical theories that aimed at proximate causes explaining why ice moved the way it did, they were fine with historical theories that framed large-scale movements of ice on geological time scales.[94]

The glacier theory framed the research agendas of most late nineteenth-century glacialists. Yet, by the end of the century there was reason to suspect that the history of glacier fluctuations was not a straightforward long retreat. The widespread, continuous recession that a retreating ice age would have demanded was being undermined by observations from the western Alps. After 1885 some Alpine glaciers, notably those near Mont Blanc, appeared to be advancing, after having been in states of retreat and then equilibrium. These fluctuations, which seemed consistent throughout the region, led some to surmise that glaciers fluctuated periodically, expanding and shrinking in decadal patterns. In his address to the commission in 1900, then president Eduard Richter (a geographer, not to be confused with Charles Richter of seismic scale fame) interpreted what appeared to be regular fluctuations of European glaciers in terms of Brückner cycles: thirty-five-year revolutions of cold, wet climate alternated with warm, dry conditions.[95] The former was thought to be conducive to glacier growth, while the latter, their retreat. Whether or not the world's glaciers were subject to periodic Brückner cycles became a major concern for the commission's members, as is suggested by the title of the introduction to its first publication: "Les Variations Périodiques des Glaciers, Discours Préliminaires" (Periodic variations of glaciers, a preliminary discourse). Engagement with the commission shifted the rationale behind the Vauxes' photographs, from tracking the demise of the Ice Age to determining whether or not glaciers in the Rockies were subject to the cycles of ebb and flow attributed to Alpine ones. Periodicity framed research and gave researchers a hypothesis to test. It also gave a semblance of predictability Europeans could use to argue for government support: if fluctuations were periodic, knowing their cycles could help prevent and mitigate damage from glacier surges and outburst floods that threatened citizens and local economies. Swiss, French, and Swedish glacier studies received government support.[96]

By 1899, contributions that challenged theories of periodicity began to trickle in from glacialists around the world. In addition to those from the Canadian Rockies, observations of Arctic and Alaskan glaciers and others in the American Rockies grew in frequency and detail. Reports came from Greenland, Scandinavia, Russia, the Caucasus, and Asia. (There were few written about South American glaciers, and Antarctica's glaciers would have to wait another decade before being subjected to the cameras and measuring sticks of men outfitted in fur and waxed canvas.) Many of these contributions reported behaviors out of sync with those observed in the western Alps. The Vauxes' work was part of this influx of new data. It established a general trend of glacier recession in the Canadian Rockies and Selkirks that did not correspond with the Alps' advance nor Brückner's cycles. "The average of all the movements of the glaciers of this region," they wrote in 1899, "has been a marked recession." In subsequent years this trend held for all but the Wenkchemna Glacier near Moraine Lake, which they suspected (rightly) was fed by avalanches from above. In 1906, after nearly twenty years of study, the Vauxes reached the conclusion that "in common with almost all glaciers throughout the world it is found that [those of the Canadian Rockies] are receding."[97]

Through the *Commission Internationale des Glaciers*, the Vauxes, and later Wheeler's repeat glacier photographs and measurements contributed to a growing sense that, overall, glaciers were retreating. By 1907, the advance in the Alps was over, only twenty-seven years after it was first observed (and so not aligned with Brückner's cycles), and signs of widespread recession from around the globe were difficult to ignore. American geophysicist Harry Fielding Reid's summaries of commission reports attest to this growing realization amongst glacialists. In 1908 he stated: "a glance at these reports will show that glaciers in all parts of the world are now retreating."[98]

Glaciers all over the world appeared to be withdrawing. Yet this apparently global recession was not a cause for alarm because it was interpreted as a natural consequence of retreating ice ages. Glacialists saw living glaciers as vestiges of a frozen past whose "natural fate" was, like the horse before the steam engine or the "Indian" before the settler, to withdraw. In all three cases, they were wrong.

Winding Down

At the turn of the twentieth century, photographs of receding termini were seen as evidence of either periodic fluctuations or ongoing ice age retreat. Understood either way, the photographs made visible that which was invisible to the unaided eye. Repeat photography in essence edits landscape change as one might splice a film reel: removing intermediate material to juxtapose moments of high drama. In this manner they create narratives of change over time. Repeat photographs "pictured the invisible" by seeming to compress time, excising and preserving distinct moments for comparison. These moments served as "pivot points" for "unpictured" processes of transfiguration. Indeed, what is not shown but merely implied is what makes repeat photography so compelling. Repeat photographs, as theorists of the genre have observed, "derive their power from a necessary reliance on the viewer's imagination of what happens outside the photographic frame."[99] In this regard they functioned similarly to the time-lapse photography pioneered by Eadweard Muybridge and Étienne-Jules Marey decades earlier, which had broken down fluid animal motion into discreet poses. Both repeat and time-lapse photography brought to light what cultural critic Walter Benjamin called the "optical unconscious": the aspects of the world that go undetected by the human eye.[100] Yet Muybridge's and Marey's work did so by seemingly extending time, revealing the discreet instances that comprise what is perceived as undivided motion. Repeat photographs, by contrast, seem to compress time and disclose the effects of motion too subtle to be detected by the eye alone. In either case, photographs reveal what we typically cannot see. When dealing with undetectably slow-moving glaciers, this is what matters.

Repeat photographs served as both revelations and mnemonic aids. They helped glacialists recall what a landscape looked like before and thereby made apparent the changes that led to the present. They rendered discernible the effects of imperceptibly slow geological processes. They also made visible hidden geographical trends. Perceiving these processes across regions through the collection and comparison of photographs helped glacialists establish hypotheses about glacier reactions to climate. While their observational scope fell woefully short of global, the efforts of the Commission des Glaciers provided a template for later monitoring efforts, such as ones made by the American Geophysical Union's Committee on Glaciers.

By the 1940s much of the infrastructure and many of the people associated with glacier naturalism were gone. The commission was a casualty of the First World War. Glacier House closed its doors in 1925. William Vaux was dead, his brother was occupied with business in Philadelphia, and his sister was busy entertaining elite Quakers in the nation's capital. The Depression brought financial hardship to the ACC, and basic concerns about membership and income took precedence over glacier work. Arthur Wheeler died in 1945. In the decades after the war, the club transformed from an omnibus organization that deemed scientific work as an integral function of an alpine association into a club devoted to the high-flying exploits of rock climbers and technical alpinists.[101] The patriotically inflected marriage of glacier study and mountain recreation that was institutionalized in the early Alpine Club of Canada was dead (at least in that configuration). The unique confluence of ideology, rational recreation, mountain topography, railway technology, and corporate support that gave rise to glacier naturalism in the Canadian Rockies had disappeared.

Yet before the threads comprising glacier naturalism had unraveled, it managed to produce visuals that did a lot of different work. Glacialists' repeat photographs of Canadian glaciers served as scientific evidence for competing theories of geological change, and in this capacity circulated through regional, national, and international networks. The photographs also contributed to imaginings of "wild" glaciers and rendering invisible the conditions that made their production possible. Glacialists benefited from and contributed to the capitalist and colonial reimagining and rearrangement of the Rocky Mountains at the expense of lay workers and Indigenous people. In their assumptions about the Rockies as their own special place, their symbiotic relationship with the CPR, and their contribution to Wheeler's and Parker's patriotic alpine club, the Vauxes helped reinforce the idea that the Rockies were a wilderness playground for bourgeois tourists to freely capture by camera and alpenstock. The format of their photographs determined by their focus on termini fluctuations framed glaciers as wilderness icons: sublime landscapes that stood for the idea of wilderness. Dominating the frame, they appear as portraits of ice against a backdrop of empty mountain lands. The photographs were scientific evidence that said a lot more than just "the ice ages are receding."

2

TRANSITIONS / *The Limits of Photography*

There are two glaciers in North America which are so frequently visited
that we should be able to keep a pretty complete record of their changes.
These are the Illecellewaet [*sic*], in British Columbia, and the Muir.

HARRY FIELDING REID / "Glacier Bay and Its Glaciers," 1896

At the turn of the twentieth century, geophysicist Harry Fielding Reid believed
that two North American ice streams were important for glacier study: the Illecillewaet, made famous by the Vauxes, and the Muir in Glacier Bay, Alaska
(fig. 9). Reid's inclusion of Muir indicates important changes and a new era in
the geography, methodology, and research priorities of North American glacier
study during the decades bracketing the turn of the century. Between 1910 and the
mid-1940s, glacier naturalism gave way to a new regime of research: geophysical
glaciology, which was characterized by quantitative methods, new technologies,
a new geography of fieldwork, and new priorities in research questions. On the
North American continent this transition was spurred by a focus on Alaska's ice
and the influx of novel methods and research agendas from Europe. The result
was a regime of scientific glaciology in which photography was demoted in importance, but this was only a hiccup in repeat photography's preeminence in the
history of glacier study.

    The details of geophysical glaciology are the topic of the next chapter. Grasping
the processes undergirding the shift in method and evidence that led to geophysical

FIGURE 9 Muir Glacier, 1890, taken by Harry Fielding Reid. Glacier Photograph Collection, National Snow and Ice Data Center, Boulder, Colorado.

glaciology is the business here. The colossal ice streams of southeastern Alaska harbor many stories. Behind repeat photographs of Alaskan glaciers lie a complex history of their making, which is ultimately also a history of their limitations. Understanding how this can be true requires delving into the conditions of glacier study as the nineteenth century gave way to the twentieth. The physical geography of southeastern Alaska is sculpted by water and weather. The combination of water and the boats needed to navigate it was conducive to the gathering of repeat photographs. These photographs, ironically, revealed conundrums that pointed to the limitations of repeat photography. Our guide through these transitions will be William O. Field, a bridge figure whose career spanned the transition from glacier naturalism and geophysical glaciology and research traditions on both sides of the North Atlantic. A prolific photographer of ice, Field's work helped reveal repeat photography's limits in telling the stories of the ice.

Alaska's Ice

Roughly 5 percent of Alaska is encrusted in ice.[1] Valley and cirque glaciers nestle among the peaks of the Brooks and Alaskan Ranges. Aleutian volcanoes sport permanent caps of ice that appear in satellite images like pearls strung across the Bering Sea. The most renowned and spectacular ice streams are found on Alaska's eastern coast. Along the Gulf of Alaska, moisture-burdened ocean winds collide with the sheer walls of the Kenai, Chugach, Coastal, and Saint Elias Ranges, depositing heaps of precipitation. Here, tidewater (ocean-terminating) glaciers crawl down from sources high in the mountains, eventually to have their termini lapped by salt water. These colossal ice beasts captured the imagination of the first glacialists to the region who, like their peers in the Canadian Rockies, saw these behemoths as living fossils. Tidewater glaciers, however, gave even more telling clues to the ice ages puzzle, for they demonstrated how perennial ice could exist at low elevations in nonpolar regions, something that would have been the norm in any ice age worthy of the name.

Alaska's tidewater glaciers are special. They are subject to factors that most mountain glaciers are not: the action of tides; regional volcanism; and rapid erosion rates. The weather is highly variable, to put it mildly—many field researchers would not employ such a disinterested term. Extreme differences in elevation between sea and summit generate variations in air temperature that drive rapid energy exchanges between glacier surfaces and the surrounding atmosphere.[2] The rhythmic swishing of tides along the glaciers' underbellies further complicates the dynamics of these already complex ice streams. Overall, the topography and climate of coastal Alaska create conditions for active and changeable glaciers. Many are prone to rapid surges and recessions. For instance, the Columbia Glacier in Prince William Sound began a dramatic recession in the early 1980s, retreating 15 kilometers by 2005.[3] Similarly, in 1937 the Black Rapids Glacier plowed across Alaska's Delta River Valley at a hasty rate of 8 meters per day, and in 1956–57 Henteel No' Loo' (the Muldrow Glacier), which flows off of Denali's broad shoulders, quietly slinked more than 6 kilometers under the cover of winter's snows.[4] While

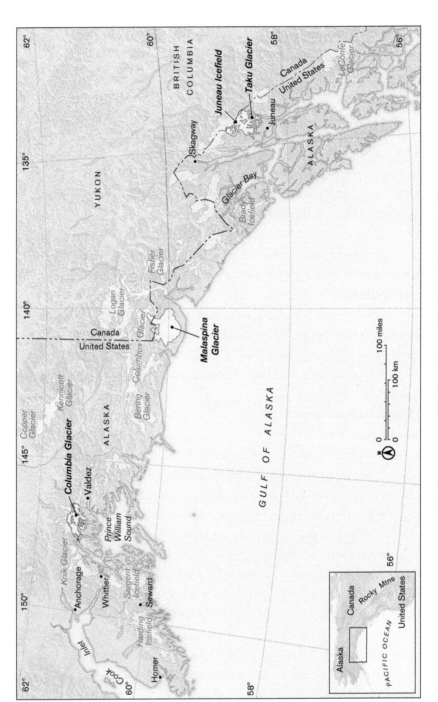

MAP 2 Southeastern coastal Alaska and some of its glaciers. Map by Pease Press.

many glaciers in the state have thinned and retreated since the eighteenth century, several others have advanced. In some cases, glaciers fed by the same névé display opposite trends: one advances while the other retreats. For early glacialists from the Lower 48, Alaska's ice seemed full of surprises. Certainly, it presented puzzles.

Coastal Alaska posed different conditions for doing fieldwork and its glaciers posed different questions to answer, than did the Canadian Rockies and their smaller mountain glaciers. Marine-glacial geography and the transportation technology it required—boats—bred a combination of environmental demands and technological answers that redefined glacier fieldwork, hoped-for outcomes, and the very things studied: the glaciers. How Alaska's glaciers were understood and studied cannot be grasped without attending to both the features of the landscapes in which they resided *and* the technologies used to access and engage with them.[5] The Canadian Rockies presented a situation in which rugged mountain topography and the rigidity of the railway limited the number of sites glacialists could visit in a single season. The railway was an inflexible backbone that, along with imposing alpine terrain, constrained the number and distribution of glaciers that glacialists could conveniently visit. By contrast, in southeastern Alaska, marine conditions and motor-powered boats made it possible to cover more geographical area and compare data from more termini. Less a spinal cord with limited peripheral nerves, the patterns traced by these seagoing glacialists resembled the tangled lines of a cat's cradle. Their more robust datasets revealed perplexing patterns in the behaviors of dynamic tidewater ice streams, challenging them to rethink simple explanations of glacier-climate interactions and whether photography was really the best tool for understanding them.

Listening for Other Stories

While glacialists may have needed photographs to show them that Alaska's glaciers are complex individuals, we can look to the cultural geography that also shapes this region for the same lesson.[6] Before there were photographers on Alaska's glaciers there were others who knew the glaciarized coasts intimately but whose knowledge was largely ignored. Indigenous knowledge of Alaska's glaciers is not the focus here. However, even a brief introduction to some of that knowledge reveals that glacier naturalism in Alaska followed similar colonial patterns as what took place in the Canadian Rockies. It, too, was founded upon assumptions of wilderness and discovery that, in part through repeat photographs, shaped subsequent understandings of the region and its glaciers.

Glacialists were not the first to notice the volatility of Alaska's ice. There is a rich history of people knowing the ice streams prior to the arrival of photographing glacialists. What is today referred to as southeastern Alaska is the traditional homelands of the Lingít Aaní (Tlingit), Dene, Eyak, Sugpiag, and Alutiiq Peoples. They knew well the extraordinary character of the glaciers among which they had lived since time immemorial. For thousands of years the ice served as trade highways between the coast and the interior, and in some cases as the geographical bottlenecks through which the coastal Tlingit monopolized trade with Russian, British, Spanish, and (later) American traders. The glacier stories of elders Annie Ned, Kitty Smith, and Angela Sidney, written down in the 1970s and 1980s in collaboration with anthropologist Julie Cruikshank, feature glaciers as members of a social order that embraced humans and other-than-humans. The stories told and retold by these women recall the Little Ice Age and contain codes of conduct for when interacting with glaciers. These included prescriptions for preparing food (boil, do not fry; be careful not to spill grease lest it rouse the glacier to surge!), lessons of humility (never taunt a glacier!), and prudent travel methods (beware crevasses!). These rich pulsating stories resist being reduced to a single exportable meaning, but they do all emphasize the dynamism and unpredictability of Alaska's glaciers, portraying them as "shape-shifters of magnificent power" that demand respect, abhor hubris, and dispense swift retribution.[7]

This dynamism is captured in a story relayed by Chookaneidí (Tcukanadi) clan elder Amy Marvin. She told of an ancestral village at Glacier Bay where the Tcukanadi and three other clans enjoyed an abundance of salmon. Their occupation of this place came to an end when Kaasteen, a young girl weary of her menstrual confinement, whistled to the nearby glacier through charmed fish bones, beckoning the ice to descend. The village took council and decided they would abandon their homes but that Kaasteen must remain (in some versions Kaasteen's grandmother elects to stay in her stead). Clan members waited until the very end to depart, watching from their canoes while icy water flooded the village and the house where Kaasteen (or her grandmother) lived "slid downward to the bottom of the sea before their eyes." Songs were composed so that the event would not be forgotten, before the people separated into different groups, one of them founding the village of Hoonah, present-day home of the Tcukanadi clan, where Amy Marvin related it.[8] Stories like this belie a knowledge of ice that is natural *and* social. Indeed, such stories push against these categories as a framework for conceptualizing the world.[9] They highlight the prickliness of glaciers and their propensity to surge without warning. Those who would speak knowledgeably about them, such stories warn, must observe, listen, and respect.

Some present-day scientists have learned that their work is improved by taking these lessons to heart and collaborating with Indigenous knowledge-holders.[10] Most turn-of-the-twentieth-century glacialists, however, heeded them not and cultivated a form of colonial ignorance that is best captured in Harry Fielding Reid's claim that "Alaska has no history except a geological history."[11] While content to consider the century-old reports of European navigators as scientific evidence, glacialists generally treated Tlingit knowledge as unreliable. Some regarded it as mere superstition.[12] As in the Canadian Rockies, where glacier naturalism contributed to colonial reimaginings of mountains, Alaskan glacialists and their forebears conceptually and visually extracted the glaciers they studied from the conditions of their work. Repeat glacier photographs erased Indigenous presence and knowledge and populated visions of the land with ideas of wilderness. This was not because they were unaware of Indigenous knowledge. On the contrary: Native men captained their boats, provided lodging in the back country, guided them through forest and muskeg, and advised them on the previous locations of glacier fronts and margins. A variant of the Kaasteen story even appears scribbled in the field diary our protagonist, William Field.[13] Yet, the geographical and glaciological knowledge required to perform these tasks and tell these stories went uncredited in glacialists' papers and books. Historian Peder Roberts has suggested that relationships among glacier researchers and Indigenous people in Alaska were likely more extensive and mutually respectful than is implied by written records. But personal relations, he recognizes, don't undo structural erasure.[14] Indigenous knowledge-holders were treated as (mere) guides, day laborers, and makers of Christmas-gift curios, not as knowers, and they are virtually absent from the writings of the men studying Alaska's glaciers. This way of operating made it possible for Europeans and Americans to claim "discovery" of the glaciers, redefining the land in terms of wilderness. Photographs of glaciers as empty landscapes devoid of human presence were one of the ways they did so. As with the Vauxes, their photographs were squarely situated within iconographies of wilderness in which epic glacier fronts came to stand for an Alaska that was empty and open for tourism and science. The beginnings of this story lie far back in the eighteenth century.

Early Encounters

Thousands of years ago, when more of the oceans were locked up in Brobdingnagian ice sheets, the Eurasian and American poles of the Bering Strait were connected by dry land. As the ice sheets retreated, spilling their meltwater into the seas, the bridge was inundated, splicing Beringia into two flanks bracketing a new strait.[15] In 1648, during the Little Ice Age, when the strait was again shallow, the Cossack Semyon Dezhnev sailed across it, returning with tales of a *bolshaya zemlya*—big land—to the east.[16] Eurasian hunters and fishers crossed in search of furs and fish. Yet these early modern intruders left little record of their encounters with the mighty glaciers on the other side of the Bering Strait. It was not until the eighteenth century (still within the bounds of the Little Ice Age) when nautical expeditions captained by Vitus Bering (1741), Aleksei Chirikov (1741), and Jean-François de Galaup, comte de Lapérouse (1786), prowled Alaska's coast seeking furs, trade routes, and geographical knowledge that Europeans recorded their experiences of the glaciers. Passing Glacier Bay in 1794, British Navy commander George Vancouver and his men beheld a landscape choked with ice, a place the Huna Tlingit called S'e Shuyee (Edge of the Glacial Silt).[17] There is no Glacier Bay showing on Vancouver's charts because at that time there was no bay. The ship's naturalist, Archibald Menzies, described what seemed to him a bleak scene dominated by "a huge mass of rugged ice" looming over "black & barren" country.[18] His observations lacked detail because the berg-infested waters threatened the ships' wooden hulls, preventing close examination. Judging by Menzies' morose tone, it seems unlikely he would have wished for a close encounter anyway.

Over the next eighty-five years the Northern Hemisphere began to emerge from the cold clutches of the Little Ice Age. As temperatures warmed and winds dampened, the ice Menzies once observed had retreated more than 20 kilometers, exposing a long, bifurcated fjord encircled by glaciers. This fjord, to which the Tlingit swiftly returned, calling it Sít Eeti Geiyí (Bay in Place of the Glacier), came to be called Glacier Bay on maps.[19] By the 1870s, at the place where Vancouver had been daunted by looming crags of ice, tourists steamed through open waters to revel in the novelty and grandeur of cerulean ice.[20] The geological opening of Glacier Bay released a profusion of words describing and celebrating the glaciers. Many of them were penned by the man for whom the bay's most famous ice stream was named: John Muir, America's "wilderness prophet."

Muir was the son of a severe Scottish minister who found his home country's kirk insufficiently pious and so moved the family to a farm near Portage, Wisconsin.[21] In 1867, when he was twenty-nine, Muir nearly lost his sight in an accident at the wheel factory where he worked. This prompted him to reevaluate his priorities and abandon wheel-making for a life as a rolling stone. "Joyful and free," he chose to "become a tramp!" Eventually, he came to the Sierra Nevada, which he deemed "mountains of light." Thenceforth Muir devoted his days to wandering in high places and writing about his adventures in florid prose that captured audiences ravenous for nature writing.[22] The Yosemite Valley, enchanting home of the Ahwahnechee people, stole his heart and anchored his wandering tendencies. Fascinated by its glacial geology (a theory for which he argued vociferously, against better-trained geologists), in 1879 he made his way to Alaska in search of ice streams that could have whittled Yosemite's horseshoe valleys and sloping granite domes.

In 1895, generations after the glacier had forced the Tcukanadi clan to abandon its ancestral village, more than a century after Bering and Chirikov had passed through the region, and a dozen years following his own first visit, Muir proclaimed in *Century Magazine* the "discovery" of Glacier Bay.[23] This was, of course, equivalent to a Tlingit person traveling to the West Highlands and announcing the "discovery" of Scottish heather. Sít Eeti Geiyí (Glacier Bay) was central to Tlingit history and life. Once the ice had begun its retreat, Tlingit people had returned to the bountiful area in search of seals. Muir knew this. He and the missionary Reverend Samuel Young were led there in October 1879 by four Indigenous men, known to history as Toyatte, Kadachan, Sitka Charley, and Stickeen John. Charley, the youngest of the group, had been to the "ice mountains" and knew the way. Arriving at the bay, they were greeted by shots from Huna sealers. But Muir only had eyes for what he described as the "lofty blue cliffs, looming up through the draggled skirts of the clouds," which "gave a tremendous impression of savage power, while the roar of the new-born icebergs thickened." Few writers have been so versed in the vernacular of the sublime as John Muir.[24]

Through his popular nature writings Muir's name became indelibly linked to Alaska's glaciers. Amid the tumbling moraines and berg-calving fronts he found a wilderness of living ice: instantiations of the geological agents responsible for his beloved Yosemite. Against prevailing scientific orthodoxy, he argued (rightly) that glaciers, not water, had shaped Yosemite. Alaska's landscape proved that glaciers of sufficient size could exist in temperate climates. It was for him a "land reborn"— still emerging from the vise-like grip of ancient ice, untainted by civilization—a wilderness perfect for rambling and for exaltation. Historian Gina Rumore has shown how Muir's vision of Alaska contributed to the creation of Glacier Bay National Monument (later Glacier Bay National Park) as a wilderness site.[25] Like the Vauxes in Banff, Muir played a foundational role in determining who would subsequently visit the region and how Americans would come to understand its value. Even before the establishment of the national monument in 1935, among the five thousand summer visitors traveling the Inside Passage in 1890 were tourists and mountaineers with a scientific interest in glaciers. Primed by Muir's writings, they too would see Alaska's glacierized coasts as a primeval wilderness harboring answers to the mystery of ancient ice sheets.

Muir is often celebrated as a father of Alaskan glaciology. Yet, although his visits to Alaska made him a connoisseur of glaciers, the 1899 Harriman Expedition showed that the rambling Scotsman was not a proper glacialist. The Harriman was a floating collection of naturalists, artists, and social elites, a group financed by railroad magnate Edward Harriman. Muir stood in opposition with what his shipmate the poet John Burroughs labeled the "fearfully and wonderfully learned" men of science, armed with their technical vocabularies and scrupulous (to Muir, tedious) methods. Their methods were those of glacier naturalism. Arriving aboard the steamships that Muir's writings had helped to promote, glacialists tackled Alaska's ice streams with their cameras, notebooks, and plane tables. On behalf of the US Geological Survey (USGS), the US Coast and Geodetic Survey, and private institutions, they documented the ice at Glacier Bay, at Lituya Bay, at Yakutat Bay, and at Prince William Sound.[26]

One of the first glacialists to organize such an expedition was George Frederick Wright, a clergyman hired by the USGS who spent a month in 1886 measuring rates of flow and daily movements of the Muir Glacier. Wright infamously used these observations to argue that there had been only one ice age; a theory, it should be noted, that aligned geology with scripture better than contemporary scientific consensus did.[27] Doubting the reliability of the iconoclastic geologist's methods (Wright surveyed using points sighted on the glacier's surface instead of fixed, easily discernible objects), Reid went to investigate the Muir Glacier for himself in 1890. Sailing on the steamer *George W. Elder*, he established a base camp near Muir's terminus, from which he and his companions made brief trips around the bay in a sixteen-foot skiff. They surveyed termini and moraines, sounded the bay, measured water temperatures, and collected samples for later analysis. As it was for glacialists in the Canadian Rockies, photography was an ineliminable technology. Reid illustrated his written descriptions of topography and the behaviors of ice with photographs documenting glacier fronts and geological features. Although he did not use repeat photographs (there were few prior images for him to re-create), he was aware of repeat photography's potential future benefits and included instructions on how to re-create his photographs, providing the locations of his stations and suggestions for how to obtain good shots.

The following year, USGS scientist Israel C. Russell brought his camera and keen interest in geomorphology to the ice around Yakutat Bay and Yas'éit'aa Shaa (Mount Saint Elias). A slight man with a mountain goat's endurance for demanding alpine travel, Russell had hastily learned photography under the tutelage of physicist Ernest Rutherford so that he might join the US Transit of Venus Expedition in 1874.[28] Fourteen years later Russell was among the scientific and political elite who met at Washington's posh Cosmos Club on a late January evening to found the National Geographic Society. Photography was, of course, a pillar tool of the young society's mission to produce and promote scientific exploration. Russell's trip to Alaska was the society's inaugural expedition, jointly funded with the USGS. Its stated motive was mapping, but all manner of natural historic observations were deemed valuable and glacial observations were specially singled out. Additionally, Russell brought personal mountaineering objectives (including a reprise of his previous unsuccessful attempt to summit Yas'éit'aa Shaa that had resulted in a six-day ordeal alone in a snow cave).[29] The following year he was back, undaunted by the drowning of six men of his party in the attempt to dock on the surf-pounded shores of Icy Bay. Much terrain had yet to be mapped, and he wanted to have another go at the mountain. Part of the scientific rationale for his trips was to determine whether Alaska's glaciers fluctuated in concert with those observed in the Alps. Should that prove to be the case, he declared, "it would not only be an interesting contribution to physical geography but have an important bearing on the study of the causes of the Glacial Epoch."[30] Like the Vauxes, his investigations were framed by theories of past global climate and by the question of glacier cycles.

Reid and Russell were followed in 1899 by the Harriman Expedition, the largest and most interdisciplinary scientific exploration of Alaska's coast to date. Named in honor of its patron, Edward Harriman, the expedition was a stress-relieving vacation, an act of philanthropic support for science and art, and a performance of wealth and status. After shepherding through a series of stressful mergers and takeovers, Harriman was ordered by his doctors to take a break from merging and acquiring. Not one for low-key holidays, he opted for an opulent journey by rail and steamship in the company of family, friends, and a handful of the nation's most prominent artists and scientists. Among the intellectual and artistic illuminati on board were Grove Karl Gilbert and fellow geologist C. Hart Merriam, photographer Edward Curtis, poet John Burroughs, and, of course, John Muir. Aboard a recommissioned *George W. Elder*, the expedition anchored at Glacier Bay, Reid Inlet, Yakutat Bay, Prince William Sound, and Cook Inlet. These stops lasted between half a day and, in the case of the Muir Glacier, four days. In referring to the expedition as "another reconnaissance" where the subtext is "*just* another reconnaissance," Gilbert voiced the frustration many of the expedition's scientists who resented having their scientific agenda play second fiddle to the whims of a railroad capitalist and an eclectic mix of passengers. The *Elder* docked according to the tastes of hunters, mountain scramblers, artists, and savants of every stripe, in addition to the mechanical needs of a large steamer. The resulting compromises left scientists like Gilbert wishing for more shore leave. Alaska's glaciers, he believed, had already been broadly surveyed by men like Reid and Russell. Now it was time for more intensive studies, using the methods Muir found so tiresome.[31]

Despite the limitations imposed by traveling on the *Elder*, Gilbert produced *Glaciers and Glaciation*, a three-hundred-page contribution to the study of Alaska's ice. Repeat photography featured prominently. Basing himself in the earlier work of Reid, Russell, and government surveys, he used photography to document changes in glacier termini. One plate, for instance, juxtaposed Russell's 1891 photographs of the Turner Glacier in Disenchantment Bay with one taken by Gilbert eight years later. Like Reid, he was mindful of photography's potential for documenting future glacier changes and made sure to provide information for repeat photographers to come: dates, locations, and instructions for finding prints of his photographs. Gilbert's and Reid's anticipation of the value of repeat photographs highlights a feature of the medium that is often underemphasized, since repeat photographs always have one foot in the past. Repeat photography is always also future-oriented: anticipating the next iteration in the series.

Mere months after the Harriman Expedition slipped through its waters, an earthquake rocked Glacier Bay, filling it with ice and bringing steamer traffic to a halt. Glacialists turned north to Prince William Sound. Among the most prominent of them were the Cornell University physical geographer Ralph Tarr and his student Lawrence Martin who, in a series of expeditions sponsored by the National Geographic Society, visited Alaska several times between 1909 and 1913. In the field they used Gilbert's photographs and reoccupied stations he had used in 1899. Their collaboration resulted in *Alaskan Glacier Studies*, a hefty tome rich with photographs detailing the glaciers of Yakutat Bay, Prince William Sound, and the Copper River. Tarr and Martin used repeat photography to document changes in vegetation, moraines, and termini positions. In some cases they highlighted the recession of ice for the reader by drawing former termini positions onto their photographs in ink, a technique that would be copied much later by the makers of *An Inconvenient Truth*.[32]

Photography was an important scientific tool for these early glacialists, just as it had been for the Vauxes. They realized its ongoing potential for documenting and revealing landscape changes and so, like Reid and Gilbert, provided information for obtaining copies of their photographs and recreating them. They dotted tidewater glacier moraines with stations for photography and surveying and documented the contours of the nodal geography for those they imagined would follow, consciously laying the foundations for photographic documentation of future changes in Alaska's glaciers.

FIGURE 10 William O. Field, circa 1956. William O. Field Papers, series 7, folder 2, box 4, Alaska & Polar Regions Archives, University of Alaska Fairbanks.

William Field's Repeat Photography

Others did follow. Prominent among them was the New York–born photographer William Bradhurst Osgood Field Jr. (1904–94) (fig. 10). "Osgood" to his college buddies and "Bill" to most, Field went to Alaska, like many of his peers, in search of mountains to climb and game to hunt. He found, instead, a world of ice, which captured his imagination and his talents. I first met Field in the space where he encountered the work of his glacialist forebears: the stacks of Widener Library, where I stumbled across his biography, *With Camera in My Hands*, cowritten with C. Suzanne Brown. I got to know him better the following summer at the Alaska and Polar Regions Archives at the University of Alaska, Fairbanks. For weeks I spent my days immersed in the papers of this diligent, considerate, gentle man, emerging only to grab a bite while gazing at the distant, white-clothed peaks of Denali Park, dreaming at night about black-and-white glacier photographs from a top bunk in the improbable lodgings of an octagonal glass greenhouse at Billie's Hostel. By the time I left Fairbanks I regretted not having met Field before his death a short while earlier. We had both been in the Canadian Rockies in 1994, he as a nonagenarian, I as a child.

Field lived through most of the twentieth century and he devoted that long life to photographing ice, honing the art of repeat glacier photography to its most comprehensive and systematic form. The most prolific North American glacier photographer of the early twentieth century, Field's collection of images eventually formed the nucleus for the more than fifteen-thousand-strong Glacier Photograph Collection at the National Snow and Ice Data Center in Boulder, Colorado.[33] Yet his long life also meant that he witnessed many changes in how glaciers were studied. Field was, thus, a transitional figure who straddled two distinct regimes: glacier naturalism and geophysical glaciology. His career in glacier photography began like that of the Vauxes, as a well-to-do tourist with an amateur interest in photography. His path soon strayed from that to one of the independent naturalist. The particularities of southeastern Alaska's glaciers, his connections to European scientists, and his personal proclivities toward systematizing led to a new perspective on glacier behaviors. This allowed him to be part of the transition to geophysical glaciology, which makes him a good (if slightly tragic) guide to the ways by which glacier photography came to reveal its own limitations.

Like the Vauxes, Field first encountered glaciers as a well-heeled visitor who was versed in the genteel hobby of landscape photography. Like the Vauxes, and North American glacier study in general, he began in the Canadian Rockies and took his first deliberate photograph of an ice tongue in 1922 when he snapped a picture of the Saskatchewan Glacier while on a family vacation. The Saskatchewan is the largest outlet glacier of the Columbia Icefield and sits 70 kilometers, as the raven flies, north of the Bow Glacier, the northernmost body of ice documented by the Vauxes. Gazing down upon the ice from a grassy shoulder jutting off Mount Athabasca's southern flank, Field could see the glacier snaking down a long valley hemmed by the massifs of Mount Saskatchewan and Mount Athabasca. At that time the glacier nearly filled the valley's length (ninety-six years later the glacier I observed from the same perspective in 2016 barely peered around the bulk of Mount Saskatchewan). Field was eighteen years old then, and his thoughts were dominated by the splendor of the scene. It was his third hunting trip to the Canadian Rockies with his father and brother. "I had no thoughts other than it was interesting and spectacular," he wrote later.[34] Like the Vauxes, his first glacier photographs were a young person's tourist mementos of a landscape perceived as wilderness. Unlike trained scientists Reid and Gilbert, Field had no intention of returning for repeat images.

The Fields were an adventurous bunch. While a student in Germany, his father ran away to join the circus and was on his way to becoming a bareback horse rider before he was apprehended and sent back to his studies. He later became a railroad engineer and met Field's mother, Lila, at a party at her parents' residence. Lila was the great-granddaughter of the railroad tycoon Cornelius "Commodore" Vanderbilt, one of the wealthiest American capitalists of the nineteenth century. Field's younger brother, Frederick, who later became an active member of the Communist Party, described their mother as "simple in an atmosphere of very considerable luxury . . . she had not a trace of vanity . . . tolerant rather than haughty, democratic rather than autocratic."[35] While we should be wary of rose tinting by a scion of wealth turned communist, Lila was at the very least an outdoorswoman who loved hunting and fishing. She purportedly held a women's record in fishing, proclaimed by the two-hundred-pound tarpon mounted proudly in the family home. With such adventuresome and well-positioned parents, Field acquired a love of travel and exploration at an early age.

Field's passion for ice, on the other hand, was slow to kindle. It began with the love of high places and mountaineering that he acquired while studying at Harvard, where he was a founding member of the Harvard Mountaineering Club (HMC). He shared this love with his brother. In 1924 they, along with Field's dormitory roommate, traveled to the Canadian Rockies in search of summits. At that time it was still a refuge of unclimbed peaks. The boys were fueled by youthful enthusiasm and a belief that their single season of guided climbing in the Alps qualified them for North American first ascents. Luckily for them they had hired Edward Feuz Jr. (son of the man who had assisted the Vauxes at Glacier House) and Joseph Biner of Zermatt, two renowned and capable guides, to lead them up South Twin, the highest unclimbed peak remaining in the Rockies. Emboldened by their success, Field made another trip the following summer, this time with the intent of locating and summiting a large peak rumored to be lurking near the headwaters of the Clearwater River in a region of the eastern Rockies that still sees relatively few visitors. He and his companions scampered among the crumbling peaks nestled in the gore point where the Siffleur and Clearwater Rivers meet, near the homes of the Wesley First Nation Band of Stoney Nakoda. "In the place of Mount Clearwater," he joyfully declared in the pages of the HMC *Bulletin*, "we had found a group consisting of at least six peaks consisting [of] 10 500 feet."[36] The ascent of South Twin and the hunt for "Mount Clearwater" marked the apogee of Field's climbing career. Still, these feats, combined with his recognized social status among the elite that then dominated mountaineering clubs, were sufficient to secure his election into the alpine clubs of America and Canada. Throughout his life he remained comfortable in mountaineering circles and used the connections they provided to facilitate his work on glaciers. The Rockies and the Alps quickened Field's love of high places and taught him the tricky arts of mountain photography and mountain climbing. Alaska allowed him to put them in service of documenting glacier change.

In 1925 something mysteriously "awoke" in him while he was steaming up the Inside Passage on the SS *Alaska* en route to hunt big game on the Kenai Peninsula. The herculean glaciers cascading into the waters off the ship's starboard side caught his attention as if for the first time. That autumn he entered his senior year of college with a new appetite for ice. Unlike the Vauxes, whose predecessors' efforts were inconstant and often poorly executed, Field had a solid foundation of glacier naturalism to build upon. Combing the shelves of Widener Library, he discovered Reid's "Glacier Bay and Its Glaciers," Tarr and Martin's *Alaskan Glacier Studies*, and Gilbert's *Glaciers and Glaciation*. Already a student of geography and geology, he consulted with his mineralogy professor, Charles Palache, a member of the Harriman Expedition, who encouraged him to reach out to the luminaries of Alaskan glacier photography.[37] The following summer, newly armed with book learning and advice from senior scientists, Field set his sights on Glacier Bay, where he hoped to repeat the photographs and measurements made by the authors of the books he loved.[38] Many years later he reflected that "at the time my plan was by no means to undertake a life's work. It was simply to try to carry out studies on the changes of the glaciers."[39]

His first excursion was a mixed affair, both in intention and outcome. Field was still a young man with an appetite for adventure. He went to Alaska with two college friends, planning to take glacier photographs and make a mountaineering reconnaissance mission to the Brady Glacier. The Brady, they hoped, set between Glacier Bay and the eastern slopes of the Fairweather Range, might yield passage up the mighty Fairweather peaks, perhaps Crillon or La Perouse.[40] At this time the highest points in North America were quickly being climbed and claimed. Those remaining in the Saint Elias Range, of which the Fairweather group were some of the highest, beckoned ambitious mountaineers. Members of the HMC and men like geographer-mountaineer Bradford Washburn and his confrere Walter Wood set their sights on the remaining unclimbed behemoths dotting the Alaska-Yukon border. Field was caught up in this peak-seeking bonanza. He made the glacier photographs of his predecessors serve two purposes: they were objects of natural history and sources of climbing information; "beta" as climbers would say today; "poop" in the vernacular Field knew.

Unfortunately for them, the lads were rebuked by the heavily crevassed Brady. Fortunately for their glacier work, this forced them to spend more time in Glacier and Lituya Bays on the boat they had chartered, the M/V *Eurus* (skippered by a colorful Hungarian whose résumé, it was said, included playing tuba for the New York Metropolitan Orchestra and serving as jailor in Juneau's penitentiary).[41] Field was greatly assisted by the Canadian guide Andy Taylor, who he had hired on the recommendation of HMC president Henry Hall. Taylor was already a local legend from his role in the first ascents of Tsalxhaan (Mount Fairweather) (4,671 m) and Mount Logan (5,959 m). Extraordinary tales of Taylor's fortitude circulated among Alaska's gossip circles, including one in which he mended his own broken leg alone in the wilderness by tying it to a tree and hauling on his leg until the bones realigned.[42] Taylor was more than just tough; he proved especially capable in the finer points of glacier photography and surveying. Field would work with him again. Together they visited over fifteen ice fronts and occupied stations previously established by the USGS, by the 1913 International Boundary Survey, and by Martin and Reid. At Muir Glacier they paid homage to Reid by reoccupying his "Camp Muir."[43] Upon returning to Massachusetts, Field wrote articles for the *Harvard Crimson* and the *Bulletin of the Appalachian Mountain Club*: his first glacier publications.[44] He had begun his life's work, though he did not know it.

It took Field some time to settle into the activity for which he would gain renown. After graduation he spent the next few years getting married, fathering children, traveling the world, and filmmaking, notably in the Caucasus.[45] It took the Depression to cement his commitment to glacier studies. The havoc and distress of the 1930s magnified the dissonance between his privileged existence and the harsh realities afflicting many folks. His brother, Fred, responded to the inequalities laid bare by the market crash of 1929 by embracing communism. The more politically moderate Bill sought a path that would lead to a "useful" life. He found it in contributing to scientific knowledge of landscape change.[46] Thus, encouraged by the earlier generation of glacialists, particularly Lawrence Martin, to whom he had been introduced as a student, Field planned a return to Alaska in 1931. Securing the help of the ever-reliable Andy Taylor, he set off for Prince William Sound with the intention of beginning of a program of systematic glacier monitoring that would connect the past to the future. Something useful.

Prince William Sound is a northern pocket of water nestled in the long crook of Alaska's coastline, approximately 650 kilometers west of Glacier Bay. It is situated in the ancestral lands of the Sugpiag, Chugach Alutiiq, Ahtna, and Eyak, and was christened for William IV, Britain's "Sailor King," in 1778, when he was but a prince.[47] The sound is large and open, with many deep fjords and bays. Port Nellie Juan, College Fjord, Unakwik Inlet, Columbia Bay, and Valdez Arm radiate off it like grasping tendrils, each a den of massive ice streams like the Columbia Glacier, and its fellows in College Fjord: the Harvard; the Bryn Mawr; the Smith; the Yale; the Wellesley; the Holyoke; and the Vassar Glaciers. The names inscribed on maps of the area evoke distinct eras in the region's colonial reimagining. Field's work was part of what we might call the "Ivy League era."

When he arrived in 1931, the last detailed observations of the glaciers were those of Tarr and Martin, taken twenty years earlier.[48] Wiser after a few close calls on the *Eurus* (including a night spent a-kilter, listing dangerously until morning, when incoming tides righted her), Field hired a larger vessel, the *Virginia*, which came with a smaller outboard motor skiff that could be rowed through shallow waters. From late August to mid-October he and Taylor photographed and mapped glaciers in Port Valdez, Unakwik and Surprise Inlets, College and Harriman Fjords, and the Copper River Valley. Almost half of their time was devoted to the immense front of Columbia, the largest glacier in Harriman Fjord. Basing themselves in an abandoned miner's cabin that looked like it had been "turned upside down by a bear," they hiked to Gilbert's 1899 stations (which Tarr and Martin had also reoccupied) and established new ones. The Columbia's surface had lowered drastically since 1910. This was made alarmingly clear by old roads that had once accessed mines by crossing the glacier. When Field and Taylor arrived, they dangled freely from cliff edges high above the Columbia's surface.[49]

The trends Field documented with his camera contributed to a growing appreciation of the localized variability of Alaska's glaciers (fig. 11). Beginning in 1910, six large and seven smaller glaciers in Prince William Sound were retreating, while two large and one small were advancing. The Valdez Glacier was retracting, the Meares advancing; the Harvard had crept forward while the nearby Bryn Mawr had drawn back. Although few of the glaciers in Harriman Fjord had changed appreciably, the Harriman had grown while the Toboggan had receded. Overall, many small glaciers appeared to be shrinking. These trends did not correspond to geographical proximity, which was odd because it was thought that since glacier fluctuations are caused by alterations in climate, glaciers in a single region, subjected to similar weather patterns, would generally fluctuate in sync. Yet Field's observations, built on those of his predecessors, showed that whether a glacier was advancing or receding often seemed irrespective of what its neighbors were doing. The advances in College and Harriman fjords were separated by a mere 51 kilometers. Yet, between them lay several retreating ice tongues. And there was a lot of temporal variability too. Since the Harriman Expedition, the ice had ebbed and flowed like fickle tides. The mighty Columbia, for instance, had advanced between 1899 and 1909, then began retreating around 1914, leaving the central

FIGURE 11 Three incongruent glaciers, St. Elias Mountains, 1972, taken by Austin Post. This image captures in one what it normally took several repeat images to establish. The leftmost glacier is advancing, the center retreating. The status of the rightmost cannot be determined from the photograph. Glacier Photograph Collection, National Snow and Ice Data Center, Boulder, Colorado.

terminus in 1931 somewhere between its 1899 and 1909 positions. Meanwhile, the western margin of the glacier seemed to have advanced.[50] Field was not the first visitor to pick up on this erratic behavior, long known to the area's Indigenous peoples. His predecessors had suspected that Alaska's glacier trends were anything but simple. Yet Field had the benefit of building on their work. Repeat photographs are evidence of change only when enough of them have been taken to illustrate it. His repeat photographs made the vacillations of Alaska's ice starkly visible.

Upon his return from Prince William Sound, Field began organizing. Ever the systematizer, he was keen to establish a centralized base from which researchers could access glacier observations. During a visit to Washington, DC, he and Martin began making plans for a collection of glacier photographs to be housed in Martin's office at the Library of Congress. With the help of geologist François Matthes, Field and Martin procured $100 of USGS money leftover from a recent expedition, which they used to buy a cabinet for storing photographs. To Field's they added Merriam's collection from the Harriman Expedition. Thus began Field's lifelong efforts to collect and centralize knowledge of North American glaciers. Encouraged by the responses he received from Martin, Matthes, and Isaiah Bowman, editor of the *Geographical Review*, he drew up plans for a program of regular glacier photography that would bring him back to Alaska every five years. In 1935 he made good on that intention, when he returned to Prince William Sound and Glacier Bay with botanist William S. Cooper.[51]

The trip resulted in significant publications for Cooper and Field, and prompted fellow mountain lover and glacier enthusiast Walter Wood to hire Field into the Department of Exploration and Field Research at the American Geographical Society (AGS).[52] No offices were available in 1940, so Field set up shop in the society's exhibition room.[53] Despite the lack of privacy, the AGS was not a bad home for Field's glacier studies. Founded in 1851 with the intention of organizing searches for Sir John Franklin's lost polar party, it was the first American not-for-profit organization devoted to the dissemination of geographical knowledge. The Department of Exploration and Field Research was founded by Walter Wood in the 1930s. Field already knew him through the exclusive Explorer's Club in New York City. The two shared a love of Alaska's glaciers and a desire to centralize knowledge of them. Field developed a three-pronged program built around repeat photographs that included field studies, collecting and organizing materials relating to glacier fluctuations, and disseminating information about glaciers to the reading public.[54] He took sole custody of the collection that was housed in Martin's Washington office and had it moved to New York. Using the AGS's contacts with other societies and geographers around the world, Field transformed his office into a magnet for glacier photographs, maps, and data. The US National Park Service, the Mazamas Mountaineering Club, the USGS, and private individuals all contributed to the collection during its early years. At times Field could barely keep up with the correspondence providing and requesting information.

The path traced by his early career thus led him from independently financed expeditions that aimed to both study glaciers and ascend mountains to an institutionalized effort to collect and compare repeat glacier photographs, maps, and data. The photographs of Alaska's glaciers amassed by Field, and the international connections afforded by his new situation, soon gave him the resources for thinking differently about glacier behavior. This in turn led to an appreciation of the limits of the very method that made the puzzles visible.

The Limitations of Repeat Photography

Photographs, observed François Matthes, "taken from definitely located and properly described points, are extremely valuable, as they record many changes in the form and appearance of a glacier which are not readily noticed by the eye or described in words."[55] Repeat photographs and surveys were at the heart of Field's glacier work. His approach was that of glacier naturalism, rooted in the geological and geographical origins of glacier study and pioneered by glacialists like the Vauxes, Martin Lawrence, and Grove Karl Gilbert. Like his forebears, Field treated glaciers as natural archives of landscape transformation and used photography to make their crawling, invisibly slow changes visible. In addition to revealing changes in a single glacier over time, repeat photographs revealed hidden geographical trends, such as those detected in Prince William Sound in 1931. Repeat photographs augmented vision, divulging what was otherwise undetectable. They did so by revealing and preserving. Typically, glaciers do not move quickly enough for their motion to be detectable by unaided human vision. Repeat photographs could reveal *because* they preserved. As Gilbert pointed out in 1903, photographs operated as mnemonic devices by helping glacialists accurately recall what landscapes previously look like.[56] They augmented sight because they augmented memory. By making the AGS a hub for glacier photographs, Field became a glacier archivist, a keeper of glaciological memory. And his AGS collection continued to serve as a central memory bank for glacier researchers, until it joined the World Data Center (A) for Glaciology in 1958.

Yet, ironically, the photographs that Field amassed and safeguarded helped usher in their own demise. The incongruent patterns of retreat and advance captured by Field, Reid, and other glacialists indicated more convoluted relations between glaciers and climate than were initially suspected. The Vauxes had viewed the retreat of Rocky Mountain glaciers simply part of the process by which the ice ages had retreated. So, too, had John Muir regarded the glaciers of Alaska: they were remnants of an ancient army of ice now retreating from its invasion of lower latitudes and elevations. Such broad theorizations were underpinned by observations of relatively congruent glacier behaviors. The wildly different trends of geographically proximate glaciers in Alaska as revealed by repeat photography did not, ultimately, lend themselves to such theories. Nor did they support theories of regular, predictable cycles. Instead, they pointed to a need for new kinds of research.

Visual studies scholars have noted that repeat photographs rely on the work of the imagination to fill in the process between two pictured states.[57] What, exactly, happens between any two moments in time is underdetermined by simply observing those two points. Just as a route between two points on a map is not immediately implied by their positions, we need to know more about what is going on in-between. In the case of the map, there may be topographical features that restrict the possible routes one could trace. Analogously, for repeat glacier photographs there may be several reasons for glaciers to change shape and extend or retract over time. Repeat photographs of Alaska's glaciers ultimately failed to tell the whole story. They could not say why one glacier was advancing while its neighbor was retreating. Nor could they, on their own, relate glacier fluctuations to climatic patterns. More and improved regional weather observations were needed. It was "idle to speculate," Field suggested, about which climatic factors were at play without having access to better weather data.[58] Although he never abandoned repeat photography as a means for studying glaciers, as he became better acquainted with the movements of Alaska's glaciers he came to recognize the method's limitations. More than cameras were needed to explain the antics of Alaska's ice. The photographs strongly suggested that glacialists lacked the information to characterize the processes occurring in-between the repeat images.

This information deficit was, in part, a matter of geography. Photographs focused on termini fluctuations. These fluctuations, as glacialists well knew, were the outcome of prior affairs higher up on the névé, where snow accumulated, settled, and began its transformation into ice. The process does not happen uniformly; it is subject to variations in the névé's topographical situation, localized micro weather patterns, and the ice's subsequent journey over fluctuating bedrock and winnowing valleys. Resulting alterations in the glacier's structure flow downward commensurate with the gravity-compelled movement of the ice. This means that termini photographs are limited spatially and temporally. They document only the state of the terminus, not the state of the overall glacier; they also capture only the result of past events, not a real-time state. Without knowing the mechanisms by which upslope changes translate to terminus fluctuations, or the lag time involved in transmitting those changes, termini observations are of limited value for assessing the overall present status of a glacier or for explaining why one glacier is thrusting forward while others nearby are in retreat.

How, exactly, glaciers respond to current climatic conditions became a compelling and focused concern for Field during his years at the AGS. In this regard he was a boundary figure. Despite his commitment to older photographic methods, he was also part of a vanguard of change sweeping American glaciology after the Second World War. This change was not altogether homegrown. The techniques and questions applied to North America's glaciers when the war ended were shaped by interwar innovations developed on Europe's glaciers, particularly those of circumpolar Scandinavia. The problem of photography's limitations was homegrown, the solution was largely imported, and Field was a key point of the new method's introduction.

Hans W. Ahlmann's Geophysical Glaciology

In 1945 British geologist and mountaineer Noel Odell drew attention to the arrival of a new mode of glacier research—"in vogue of recent years"—by distinguishing between glaciology and glacial geology.[59] Glacial geologists concerned themselves with the effects of glaciers upon landscapes, while practitioners of the new glaciology studied the physical constitution of ice. In drawing this line Odell was giving new meaning to old words. Since the nineteenth century, "glaciology" and "glaciologist" had referred to the study of glaciers in general (the latter being synonymous with "glacialist"). Yet the new, narrower meaning of the terms differentiated between the study of how glaciers sculpt their environs over time versus the study of the physical structures and processes that explain how ice behaves. Glaciology was considered a geophysical endeavor, meaning simply that it aimed to study glaciers as a subject matter in the physics of the Earth, joining growing bodies of work in seismography, electromagnetism, and gravity measurements, among other fields.

Sensitivity to these research agendas as distinct projects was prompted by the observations of turn-of-the-century glacialists. Alaska glacialists like Field, as well as members of the Commission Internationale des Glacieres, were noticing anomalies in glacier fluctuations that were too discordant to be explained by dominant theories of ice age retreat or regular cycles of growth and decay. This observation pushed them to seek more precise explanations of why glaciers moved and fluctuated as they did. It was no longer enough to know that Alaska's glaciers behaved irregularly. Glaciologists wanted the *exact* reasons why. Like many biologists of this time, they turned to physics as a model for exactitude in science. Just as biologists "looked to physicists and chemists for models of how scientific investigations should be carried out," so too did glaciologists appropriate and emulate the agendas and methods of the physical sciences.[60] "Glaciology" came to mean the physical study of ice as a branch of geophysics.

European glaciologists led the way. Since the early decades of the twentieth century, German, Swiss, and French glacier researchers had been moving away from the natural-historical, expeditionary approach of glacier naturalism and toward one that was on-site, long-term, and more technical. One of the most influential scientists leading this transition was the Swedish geographer and glaciologist Hans Wilhelmsson Ahlmann. Historian Sverker Sörlin characterizes Ahlmann as a science diplomat: a trepidatious, even insecure and anxious man who learned to move fluidly within and across international circles, becoming a world authority on glaciers, polar climate, and climatic change.[61] Ahlmann's ideas about how to study glaciers were formative in North America, in large part due to his influence on Bill Field. Ahlmann began his studies in 1918 as a geomorphologist (one who studies the processes that form the Earth's surfaces, such as erosion and sedimentation) under the tutelage of Gerard De Geer. De Geer was known for developing the technique of counting varves. Varves are very thin layers of geological strata that are laid down every year by sedimentation processes. Counting them was a tedious but accurate way of calculating from the geological record the passage of time.[62] Ahlmann shared De Geer's commitment to exactitude and quantitative methods. Although he came to glaciers wishing to know how they shaped topographies and recorded climates (questions for a glacial geologist, according to

Odell), he soon turned his quantitatively inclined mind to the study of glaciers themselves, attending to the physical processes by which water accumulated, moved through them, and was lost. Ahlmann saw himself as part of a revolution in European polar science in which the naturalist-explorers ceded power to the instrument-wielding geophysicists.

Earlier polar research, Ahlmann contended, conflated research with exploration and national heroics, making the former the handmaiden of the latter two. Men like the famous Norwegian polar explorer Roald Amundsen, who led the first party to reach the south pole, exemplified the conflation Ahlmann described. Amundsen's polar sojourns were motivated by the goal of attaining a particular geographical point, something he was very good at. Any measurements or observations collected along the way were (according to Ahlmann) mere afterthoughts. Moreover, the heroic and nationalist rhetoric underpinning expeditions like Amundsen's and those of his rival, Robert Falcon Scott, were politically problematic. During the interwar years Ahlmann advanced an ideal of polar science that downplayed international competition and instead advocated cosmopolitan cooperation for the greater good of knowledge production.[63]

Ahlmann believed that the new age of polar exploration had been inaugurated in 1930 with Alfred Wegener's Eismitte ("middle ice") Expedition, the first to overwinter amid the brutal conditions at the heart of Kalaallit Nunaat (Greenland). Wegener was Germany's most renowned polar explorer and key developer of the theory of continental drift; Eismitte was his last expedition. After a late-season supply run, Wegener perished on the return journey. The three men who remained huddled in the snow cave *cum* research station—Johannes Georgi, Ernst Sörge, and Fritz Lowe—spent the long dark winter collecting meteorological and glaciological data in situ. This, for Ahlmann, was a change in the game. Reaching the middle of the ice sheet was only the first step. The real work took place with a stack of finicky instruments while locked in place by cold, snow, and the long winter night.

In reality, the expedition was a mixture of old and new. Drawing a clear line between a "heroic age" and a "scientific age" of polar exploration is easier said than done. As historians have rightly observed, "scientifically-motivated exploration extended backwards well before 1900, while heroic exploration traditionally associated with the Victorian era extended well into the twentieth."[64] Nevertheless, for

Ahlmann, Eismitte was a break from the past, to be hailed as an achievement of a "modern," "science-first" approach that was driven not by nationalist vanity but by a cosmopolitan mission to contribute to a growing stock of human knowledge. It taught him that much could be achieved when researchers were camped out on a glacier with a few instruments and a good deal of patience. Following his mentor De Geer and the example of the Eismitte scientists, Ahlmann embraced what Sörlin has characterized as a "culture of precision," stressing stringent exactitude in the field. "It is the exact sciences," Ahlmann declared in 1931, "that come to the fore [in modern polar research]—everything is geared towards maximum precision and completeness."[65] Such was his ideal for glaciology.

Through the 1920s and 1930s he put this ideal into practice on Norway's Jotunheim Glacier and later on the glaciers in Sweden, Iceland, and Kalaallit Nunaat (Greenland). During this time he developed his most notable contribution to glacier science, what he called "regimen studies" and which are today known as glaciological mass balance studies.[66] They were a way of measuring the making and unmaking of glaciers. A glacier's regimen is the net change in water volume experienced by an entire glacier over the course of a year. It is calculated by determining the amount of water gained during the accumulation season (typically the winter) and adding that gain to the loss experienced during the ablation season (usually summer). During his winter at Eismitte, Sörge had learned that annual layers could be discretely identified by the different densities of the snowpack. Winter accumulation, then, could be found by counting layers in the walls of snow pits (a technique akin to De Geer's counting of varves) and calculating the average amount of water in them. At the time, ablation was measured using ablatographs—instruments that recorded the movement of a float resting on snow as the glacier's surface lowered—and graduated stakes installed in spring: as snow melted or evaporated, the lowering could be read off the stake (fig. 12). Rates of ablation and accumulation were documented at multiple sites and their averages were calculated and multiplied by the glacier's area, to obtain a result for the whole.

Calculating a glacier's regimen in this way was scientifically superior to documenting termini fluctuations. Termini fluctuations documented changes in only one part of the glacier, and the ability of the terminus to represent the entire glacier was a dubitable proposition. If one wished to understand how glaciers gained and lost ice, regimen studies offered a more direct and more immediate measure,

FIGURE 12 Ablatograph. Photograph from Hans Ahlmann Collection:
Album 55: Norwegian-Swedish Spitsbergen Expedition. Center for History
of Sciences, The Royal Swedish Academy of Sciences.

a quantitative (as opposed to photographic) snapshot of overall glacier health.
They made it possible, at least in theory, to determine if a glacier as a whole was
gaining or losing mass and whether the terminus fluctuations reflected this state.
Ahlmann's regimen studies brought glaciologists nearer to understanding the
source of termini fluctuations and how glaciers responded to climate. Glaciologists, he opined, knew "very little about the meteorological reasons of [glaciers']
existence and variations in size, about their structure, movement, and other features."[67] Remedying this defect required attending to the meteorological factors
impinging directly on the glacier's surface—the proximate causes for a glacier's
gaining or losing of mass. The reasons behind glacier expansion and contraction
were to be found in the interactions between weather and water in this superficial
zone, hovering just above and below the border between ice and air. What effect
did wind velocity have on ablation rate? How was surface evaporation affected
by radiation, humidity, and ambient air temperature? Until such questions could

be answered, he asserted, glaciologists did not truly understand why glaciers grew and shrank. Nor could glaciers be "utilized as the climatographical registrars" they really were.[68]

Ahlmann used regimen studies and close inspection of the micrometeorology at the glacier's surface to establish what Sörlin has termed a "microgeography of authority." This was an authority that leveraged place and precision. The place was the névé, the upslope area of snow accumulation and the source of glacier movement. The precision was supplied by instruments capable of measuring wind velocity, ambient moisture, the height of the snow surface, and solar radiation. Precision remade place. The glaciers of glacier naturalism, understood through repeat photographs and terminus surveys, were objects of natural history to be collected and compared for patterns. They were defined predominantly by their termini, the main sites of interaction between glaciers and researchers. On the other hand, glaciers punctuated by ablatographs, snow stakes, steel tubes, anemometers (measuring wind velocity), thermometers, and housing for both instruments and researchers were a new thing. Royal Geographical Society member Gordon Manley deemed that Ahlmann's protracted, on-site fieldwork using specialized instruments ushered in a new "instrumentalized era" of glaciology.[69] The "instrumentalized glacier" was a new scientific object fashioned by populating a glacier's névé with instruments and support materials (such as researchers' tents). The object was known not through a visual grasp of terminus changes. To know an instrumentalized glacier was to know about a construct of instrument-generated numbers. Its components were extracted and abstracted by instrument-wielding researchers and then reconstructed in charts, tables, diagrams, and scientific prose. The glaciers that Ahlmann probed with precise instruments high atop the névé were conceptually and practically distinct from those observed by Field with his camera at the terminus's edge. They were different objects of study for scientists, different glaciers. And they promised to reveal secrets that the camera could not disclose.

Yet, as glaciers make clear, the documentation of such alterations is never completely in human hands. The natural world pushes back, refusing to be reduced to the variables the scientists define. Pausing to consider where science fails adds nuance and humility to our histories. Ahlmann's instrumentalized glaciers were imperfect manifestations of an ideal. In practice his studies were limited by sampling techniques. Accumulation and ablation rates vary across a glacier's topography in ways that cannot be precisely accounted for by averages. Wind drifts and sheltered areas where snow gathers in greater quantities throws off measurements. Taking readings at multiple locations around a glacier accounts for some exogenous variables and makes for a better overall picture, but such efforts only mitigate the problem. They do not solve it, especially on large, complex glaciers such as Iceland's Vatnajökull, Europe's largest glacier system and the site of Ahlmann's 1936 fieldwork. To determine how much water was lost through ablation, glaciologists had to be certain they accounted for loss through streams flowing from the ice. This was rarely a simple task, and in the case of Vatnajökull it was nigh impossible in the conditions Ahlmann and his team faced that season. His expedition narrative describes repeated pummeling by the North Atlantic storms that periodically beset Iceland. Some nights researchers had to take turns continuously excavating their tents to prevent being buried alive by furious blizzards. Under such conditions, obtaining any measurements was an extraordinary feat.[70] As is often the case in field science, the scientists were not in charge.[71] Their struggle to survive on Vatnajökull, much less capture a comprehensive data set, is a humbling reminder of the dangers and difficulties of conducting fieldwork when the object of a study dwarfs the scientists and their instruments. Although Ahlmann's Vatnajökull work was later criticized for failing to account for all possible sources of water loss on the massif, it is worth remembering that an instrumentalized glacier is at best an imperfect and incomplete modification of a large, recalcitrant object of study.

Although imperfect and often conducted under trying conditions, Ahlmann's research of the 1920s and 1930s revealed a trend of circumpolar glacier recession. The Jotunheim glaciers, Svalbard's Fourteenth of July Glacier and Isachsen Plateau, Iceland's Hoeffellsjökull, and the Fröya Glacier in Kalaallit Nunaat (Greenland) were all showing negative regimens by the late 1930s. After periods of advance punctuated by the occasional retreat, glaciers in every corner of the European North Atlantic seemed to be shrinking (as the Commission Internationale des Glaciers had surmised earlier). This, Ahlmann suggested to an audience at the Royal Geographical Society in 1946, pointed to a recent climatic "improvement" or "amelioration" that could be attributed to changes in atmospheric circulation observed by society fellow Gordon Manley. According to Manley, warmer air was coming to higher latitudes for longer periods of time, rendering circumpolar climates, like that of Norway, more maritime. Ahlmann's micrometeorological work had suggested that extended periods of warm humid weather made more of a difference to a glacier's regimen than did higher winter accumulation levels. Thus, it would seem, longer stretches of mild weather would have considerable impact on the glaciers.[72] Today, "improvement" and "amelioration" in the context of global warming sounds like the work of a public relations spin doctor. For Ahlmann, who had little reason to believe the warming was due to human actions, it seemed a good thing. As one who knew well the difficulties of living and working in cold environments, and who could not have anticipated the dire effects of widespread, rapid warming, he supposed a bit of warming might improve life and agriculture in the high north.[73]

Ahlmann's influence on midcentury glaciology stretched across the globe. However, he was not the only person to advocate for new methods and a new research agenda in glaciology. There were others in Britain, Switzerland, and Russia who also pushed glacier study in new directions, particularly in the realms of laboratory and mathematical research.[74] Yet Ahlmann's influence on North American field glaciology was early and readily discernible. This was in part because he explicitly advocated for his névé-based, instrumentalized approach in international venues. He was a "science diplomat" who encouraged international scientific cooperation to address questions in the earth sciences. He deftly turned the obvious geographical limitations of his polar warming thesis into an argument for worldwide glaciological research. The extent and causes of the supposed warming trend could only be determined by looking outside the Scandinavian subarctic. "If we find in the Antarctic similar evidence of the present climatic fluctuation as has been found in other parts of the world," he speculated, then perhaps climatic amelioration was being caused by global factors, like solar cycles.[75] If not, scientists' perspective on the problem should change, suggesting that the polar warming—which should then rightly been called "Arctic warming"—was caused by regional meteorological factors. It is unsurprising that Ahlmann was a key instigator of the 1949–52 Norwegian-British-Swedish Antarctic Expedition, which aimed to discover what antipodal glaciers were doing. Writing widely in modern science's lingua franca, his articles in *Geografiska Annaler* (the English-language journal of the Swedish Society for Anthropology and Geography) and the *Journal of Glaciology*, the *Geographical Review*, the *Geographical Journal*, and the *Polar*

Record, encouraged the scientific world to take up the cause of instrumentalizing glaciers. In a 1948 issue of the *Geographical Review* he provided a template for the ideal instrumentalizable glacier: small, of simple geometry, relatively accessible to research parties, and possessing a well-documented history of fluctuation and regional weather records.[76] His work reached broad audiences, especially those with geographical training, including the AGS's new employee: Bill Field. As Field stated, glaciology proper—by which he meant glaciology in the sense articulated by Odell—began with Ahlmann's work. He was not alone in thinking this. Other North American glaciologists regarded the Swedish scientist's work as the genesis of systematic glaciological research.[77]

Let's Get Geophysical

The new priorities and practices of glacier research were reflected in the visual representations glaciologists used. Repeat termini photographs appear relatively rarely in the publications of interwar European glaciologists. Instead, readers were presented with stereograms (three-dimensional images) illustrating the orientations of snow and ice crystals, charts, graphs, tables, and plenty of equations. Ahlmann's voluminous *Geografiska Annaler* reports featured graphs of ablation and accumulation rates, with meteorological data organized in tables and charts. To be sure, photographs still appeared in these works. Men like Ahlmann were pursuing a new type of research in the field and they were aware of themselves as doing so. Documentary photographs of instruments, techniques, and personnel were common, as were maps and sometimes photographs that captured interesting geological features and formations. Repeat glacier photographs, however, were not the central type of visual evidence because terminus fluctuations were not the focus.

Ahlmann understood his approach to glaciology as a turning away from an earlier, heroic and nationalist mode of polar research. Exploration and physical exertion were, at best, secondary effects of the primary goal: detailed, long-term, in situ research into the physical attributes of snow and ice and the meteorological factors that impinged upon them. From the historian's vantage point we can be more precise about what he regarded as a transition. It was not a clean switch from the exploratory and heroic to the austerely scientific. Rather, it was a moving away from the practices and research agendas of glacier naturalism. Glacialists like the Vauxes and Field were concerned with the causes of ice ages and cyclical glacier fluctuations; they relied on repeat photography and surveys of glacier termini as their primary techniques in the field. Transferred to Alaska, early glacialists engaged with the landscape as a wilderness reborn from the grip of ice ages and sought to document this process through repeat photographs. Like their colleagues in the Rockies, they operated in a mode, exemplified by Muir, that officially discounted the rich knowledge of glaciers held by Indigenous people who lived there. Their images contributed to an iconography of Alaska's wilderness that made possible and contributed to the erasure of Indigenous presence and knowledge of the land. For glacialists, their photographs had value as future-facing material records that would help scientists collectively recall what a landscape used to look like. For us, we can see that they were also visual extractions of glaciers from the complex, historically embedded processes by which they were studied. They were also tools for visualizing coastal Alaska as a certain kind of place.

This was the tradition Field entered into and carried forward into the twentieth century. But this tradition carried the seeds of its own unmaking. The photographic record generated by glacier naturalists, made possible by tidewater glaciers and boats, produced puzzles that could not be solved by more photographs. The images of glacialists like Reid, Gilbert, and Field helped make apparent the perplexing differences in behavior of neighboring ice streams. Repeat photographs could reveal *that* adjacent glaciers were acting differently but could not provide an explanation as to *why*. Addressing that question required North American glaciologists to adopt the practices and agendas developed by interwar European scientists such as Ahlmann, who saw themselves as ushering in a new, more sober and cosmopolitan form of glacier science. Their work helped Field and his North American compatriots realize that repeat photographs could not explain glacier fluctuations. They needed on-site, protracted studies of the physics of ice, snow, and weather. North American glaciology needed to get geophysical.

3

MEASURING / *Geophysical Glaciology*

In the autumn of 1947 Bill Field and Walter Wood were talking "a good deal" about
their shared desire for "Ahlmann-type" research programs on North America's
glaciers.[1] Wood was pursuing a career at the newly established Arctic Institute of
North America (AINA) but had regular contact with Field, who, upon Wood's
departure, finally got an office. Both men felt it was unacceptable that American
glacier science lagged so far behind that of Europe. While it had been tolerable
to trail their European counterparts before the war, the heady atmosphere of
American confidence in the postwar years made that harder to bear. This was a
time of substantial development and expansion for the American earth sciences.
Oceanography, geophysics, seismology, and studies of the atmosphere and ionosphere joined geology, cartography, and meteorology as sciences of interest to
the American government and deserving of government support.[2] Glaciology
shared in this flourishing. Field and Wood soon had their wishes fulfilled. By
summer 1949 Wood was overseeing Project Snow Cornice (1948–51) with Caltech
glaciologist Robert P. Sharp and Field was running the Juneau Icefield Research
Project (JIRP) (1948–58) with mountaineer and glaciologist Maynard Malcolm
Miller. Snow Cornice and JIRP were the first sustained attempts to study North
America's glaciers in the geophysical mode described by Odell and championed
by European scientists in the 1920s and 1930s. The projects were the culmination
of the conditions already mentioned: boat-accessible tidewater glaciers, the limits
of repeat photography, and the influence of a European research agenda. They

were long-term, situated field studies that focused on the physical structures and processes of snow and ice. In their impetus and ambitions they initially resembled the work done by Ahlmann and his Old World colleagues. But the conditions—physical, political, and financial—under which geophysical glaciology was possible in North America differed substantially from those in Scandinavia and the Alps. Thus, the *practice* of geophysical glaciology on this side of the Atlantic looked very different and impacted the way photography was used on the ice.

The conundrums revealed by repeat photographs of Alaska's glaciers had made it abundantly clear to Field and his contemporaries that more than cameras and boats were needed to study how glacier ice formed and moved. But that acknowledgment was only the first step, leaving still the monumental task of figuring out how to do geophysical work in the Alaska setting. Alaska came with unique challenges that called for unique solutions. Its formidable mountain geography meant that JIRP and Snow Cornice would initially share many traits with contemporary mountaineering expeditions. But this approach would ultimately fall short of supplying the needs of long-term, névé-situated glaciology. The projects needed more funding and more material support than mountaineering expeditions; both were secured through military patronage. Thus, matters of scale, both geographical and scientific, rearranged the relations among glaciologists, their supporters, and their objects of study and impacted glaciological practice and the role of photography in the study of ice.[3] Although photographs could not answer the geophysical questions defining the new research agenda, they became useful for capturing information of interest to military planners, who sought knowledge about how to live and work in frozen environments. Inadvertently, military patronage rescued the camera from obsolescence and gave it a new part to play in the study of glaciers.

JIRP and Snow Cornice

JIRP and Snow Cornice were the first of their kind in North America. Unsurprisingly, they shared some of the features that defined earlier work. Their research agendas were rooted in European precedents. Socially they were anchored in East Coast mountaineering fraternities and Ivy League circles, which helped steer the projects toward the high, unclimbed peaks of Alaska: the same region that Field and his predecessors had explored and puzzled over decades prior. Understanding why and how the two projects evolved away from mountaineering-style expeditions illuminates the unique mode of early geophysical glaciology in North America.

North American glacier scientists had taken steps toward geophysical approaches before the war. In a speech made to the 1931 gathering of the American Geophysical Union (AGU), François Matthes argued for a new geophysical orientation in their discipline. Glaciology's proper home, he suggested, was not geology, but hydrology. As bodies of frozen water and as sites of water storage, glaciers were important elements in the hydrological cycle and North American glaciology needed to become a branch of geophysics and a subfield of hydrology. He cited the authority of the International Union of Geodesy and Geophysics, after which the AGU was modeled and to which it sought to contribute, which established its Commission on Glaciers under its hydrology section in 1927.[4] Inspired by scientists like Ahlmann in Scandinavia and Gerard Seligman and Henri Bader in the Swiss Alps, Matthes called for research that was focused on understanding the structural and dynamic—that is, geophysical—aspects of glaciers rather than querying them as remnants of the past: this was glacial geology *pace* Odell. A few American glaciologists began moving in the direction Matthes indicated. For instance, Richard Goldthwait, who would later establish the Institute of Polar Studies at Ohio State University, attempted to use seismic techniques to measure the thickness of Alaska's Crillon Glacier in 1934.[5] But such efforts were sporadic and small scale and then they were interrupted by war.

The Second World War stalled glacier study and directed relevant human resources elsewhere, with consequences for where glaciologists would turn when they reembarked on civilian life. In perilous places like the high mountains or the far north, soldiers must contend with both human and environmental antagonists. The war demonstrated that combat in cold weather environs required unique preparation. In the winter of 1939–40 the improbable success of a few vastly outnumbered Finnish soldiers against Russian invaders demonstrated to the world the importance of knowing one's environment. The Finns, traveling swiftly on skis and clad in weather-appropriate clothing that camouflaged them against the snowy backdrop, harried the road-bound, frost-bitten Russian soldiers for three months. Although the Finnish resistance eventually collapsed, it demonstrated what could be achieved when soldiers were adapted to specific environs. The Swiss, German, and Italian militaries already appreciated this lesson, as their specialized mountain units attested.[6] The Americans took note. Both European and Pacific theaters required fighters and materiel suited to cold, remote places. On the home front, Alaska's proximity to Japan—and to America's erstwhile ally, the Soviet Union—made the far north a priority from the beginning of American involvement. Japanese occupation of three Aleutian Islands in June 1942 underlined Alaska's vulnerability. The United States responded to these prompts by creating its own specialized alpine force, the Tenth Mountain Division, in 1941 and by ramping up research into equipage and logistics in cold weather environments.[7] Postwar glaciology benefited from both.

During the war, mountaineers, skiers, and glacier researchers put their high-altitude experience to work by evaluating and improving their nation's ability to fight in cold weather. Project Snow Cornice's future leaders Walter Wood and Robert Sharp served in the Army's newly established Arctic, Desert and Tropical Regions Information Center. Wood and mountaineer-geographer Bradford Washburn joined mountaineers Robert Bates, Einar Nilsson, and Terris Moore (later president of the University of Alaska) to field test equipment on Denali in 1942. Washburn returned two years later, when he and Sharp led an overland test of emergency equipment carried by US Army Air Force fliers over cold weather terrain.[8] Bill Field also enlisted. After a stint making army training films in Ohio, his expertise in extreme environments was utilized at Fairbanks' Ladd Airfield, where he worked on a film about how to survive plane crashes in cold places.[9]

Many of the recommendations and technological developments made by these climber-consultants failed to translate into improvements for soldiers in action. However, their work tilled the ground for close working relationships with the military after the war. It provided glaciologists with experiences that would help them tap into military coffers later by initiating them into the ways of martial bureaucracies and lexicons. This knowledge would serve them well later, when looking to convince army and navy patrons that their research met cold war strategic needs. Upon returning to civilian life, glaciologists were able to pick up their research where they had left off, but they were armed with new experiences.

The beginnings of the Juneau Icefield Research Project lay in the months before America's entry into the war and owe much to the tightknit social circles, and

FIGURE 13 Maynard Malcolm Miller (fourth from left) with Juneau Icefield Research Project staff, taken in 1949 by Ansel Adams who happened to be in the area at this time. From the Juneau Icefield Research Project Collections, American Geographical Society Archives, box 165, folder 37. Reproduced with permission from the American Geographical Society Library, University of Wisconsin, Milwaukee, Libraries.

outright nepotism, of Harvard-educated mountaineers. Geophysical glaciology in North America grew in soils cultivated by mountaineers. When Walter Wood left for AINA in 1947, Field was made director of the AGS's Department of Exploration and Field Research. His new administrative duties left little time for fieldwork, so he became primarily a coordinator of others. That year, on the recommendation of Bradford Washburn—mountaineer and fellow Harvard alumnus—Field hired Maynard "Mal" Miller to help him make observations near Glacier Bay and Yakutat (fig. 13). Washburn was Miller's geography instructor at Harvard and could vouch for his field experience because Miller had been one of the young men to join the Washburns on their unorthodox honeymoon the previous year. On that trip Barbara Washburn became the first woman to summit Mount Bertha (3,110 meters) and likely the first to spend her honeymoon tent-bound with five college-boy climbers and one Austrian ski instructor. Bradford Washburn was adept at combining mountaineering and geographical ambitions into the same expedition. Whether he was successful at introducing romantic ones is less certain (though, to be fair, his marriage lasted through many more expeditions).[10]

Miller embraced Washburn's practice of combining mountaineering and scientific exploration. He came to glacier study through a love of mountaineering that had been honed on the peaks of the Pacific Northwest, near his home in Tacoma, Washington. His first trip to Alaska was with the Washburns, and it was enough to hook him. The peaks of the Fairweather Range surpassed those of his home in height, drama, and challenge. Gazing upon them, Miller felt he was among rivals of the Himalaya. An ebullient person by nature, he was prone to enthusiastic amplification and tended to throw himself wholeheartedly into projects, especially mountaineering ones.[11] His climbing skills were an asset for work above the tree line. And Field quickly recognized Miller's potential as a field leader possessing infectious enthusiasm. When Field returned to the AGS after the war, he enlisted Miller again for glacier work.

Miller's desire to combine work and play would later be a point of contention between the two men, compounded by the fact that, as Miller himself confessed, he sometimes struggled with the academic aspects of glacier work. Yet this future disharmony was unforeseeable in 1946. That summer Miller was back in Alaska for a try at the southwest ridge of Yas'éit'aa Shaa (Mount Saint Elias, 5,489 meters). The mountain had not been summitted since the Duke of Abruzzi's party first ascended it in 1897. The 1946 attempt was the fulfilment of transoceanic dreams hashed out in wartime correspondence among Miller in the South Pacific and fellow HMC members William Putnam and Andrew Kauffman, both of whom were stationed in Italy with the Tenth Mountain Division. After the ascent, Miller and another team member, William Latady (also an HMC member and veteran) remained in Alaska to conduct reconnaissance work for their mutual acquaintance, Bill Field. Relying on AGS funds and $200 from the new American Alpine Club Research Fund, they made a rapid survey of more than eighty ice tongues (and, undoubtably, several future climbing projects).[12] They were particularly struck by the appearance of the Taku Glacier. The Taku is the largest outlet glacier of the Juneau Icefield, an 1,800-square-kilometer expanse of ice and snow draped over the crest of the Boundary Ranges along the present-day international border. It is stabbed by tines of submerged batholiths, the highest of which is Devils Paw (2,616 meters), which appears to punch through the snow like a clenched fist. Most of the icefield is perched above 1,525 meters and receives heavy doses of precipitation from the maritime climate to the west, making it an ideal habitat for the more than fifty glaciers that drain from it.

Unlike the other large glaciers of the Juneau Icefield (the Norris, the Twin, and the Tulsukwe), all of which were receding, the Taku was vigorously pushing forward into the Sitka spruce at its terminus. Why, Miller and Latady wondered, was the Taku (and its secondary stream, the Hole-in-the-Wall Glacier) advancing while its neighbors, fed by the same icefield, were retreating? The Taku's behavior illustrated yet again how little was understood about the factors influencing proximate but separate ice streams. Subsequent conversations with Field and a flyover in 1947 led Miller to appreciate what many glacier researchers had learned prior to the war: the ability to understand paradoxical patterns of recession and advancement requires more than just photographs. It demands detailed, physical investigations of the névé.

The Taku seemed an ideal location for Ahlmann-style geophysical glaciology. Every portion of it, from terminus to source, could be reached by skis from several approach options, either from the west via a highway and the Mendenhall or Lemon Creek Glaciers or from the east via the relatively sheltered waters of Taku Inlet. The latter had a gentle, undulating topography free of high surrounding peaks that facilitated travel, but it also meant there were no independent ice streams flowing from tall mountains to complicate flow dynamics. Moreover, termini fluctuations and regional weather were relatively well-documented, the latter dating back continuously to 1899. The Taku appeared to tick all the boxes in Ahlmann's criteria for instrumentalized glaciers: it was geometrically simple, accessible, and had decent historical records. Field believed he had a site for his Ahlmann-type glacier research station as well as a field leader in Miller. The two began developing plans for a long-term study that would tackle the upslope factors influencing the regimen and flow of the Taku.

Their plans were rooted in Ahlmann's interwar research agenda. What evolved into the Juneau Icefield Research Project began as half of a larger Glacier Studies Project that would have compared glaciers in Alaska and Patagonia.[13] Transhemispheric studies, Field urged in an early memo, were needed to determine the extent and nature of the glacier recession observed by Ahlmann—something Ahlmann himself had also said. On both sides of the Atlantic it was understood that Field and Miller were extending the Swedish research program. Valter Schytt, Ahlmann's protégé and heir to his glaciological legacy in Scandinavia, declared Field and Miller's project "the westernmost link in that chain of glaciological research that Ahlmann has organized." Field welcomed the comparison.[14] He made certain to assure potential backers that the project had been developed in collaboration with the "eminent glaciologist" Hans Ahlmann and the late François Matthes. Unlike other branches of the American earth sciences (which, historians note, often regarded European science as suspiciously left-leaning), glaciology in North America did not shrink from having European associations.[15] Referencing Ahlmann's expertise allowed Field to argue that they were closing a research gap while also claiming a measure of the foreign expertise to which they aspired.

JIRP was as an effort to bring Ahlmann's glaciology to North America using resources provided by East Coast geographical institutions and Ivy League mountaineering circles. It was fashioned in a crucible of postwar mountaineering, European research agendas, and Alaska's geography. These elements, in different proportions, also defined the inception of Project Snow Cornice, the other inaugural instance of geophysical glaciology on this continent.

Project Snow Cornice was the first long-term research endeavor taken on by the newly formed Arctic Institute of North America. AINA was a product of wartime conversations among Canadian and American academics and bureaucrats who sought to increase their nations' scientific, technical, and administrative presence in the Arctic. Its goal was to "initiate, encourage, support, and advance by financial or otherwise the objective study of Arctic conditions and problems."[16] With funding from the two federal governments and a handful of private institutions, sister offices were established in Montreal and New York; in 1947 Walter Wood was placed in charge of the latter. Project Snow Cornice was Wood's darling. It was his take on the Ahlmann-type glacier study that he and Field had discussed at length. Wood's version would combine mountaineering and scientific fieldwork in what he called "constructive mountaineering." In constructive mountaineering, technical mountaineers place their skills at the service of "mankind" by making them serve science.[17]

Like Bradford Washburn, Wood believed that mountaineering and science could be pursued harmoniously in a single expedition. Indeed, he chose Malaspina Glacier as the site for Project Snow Cornice for both its scientific and its mountaineering potential. At its back were the Saint Elias Mountains, which in 1948 were home to the tallest unclimbed peaks on the continent. Wood occasionally referred to the project as the "Mount Vancouver Expedition"—a reference to the chief mountain he hoped to summit, then the highest unclimbed peak in North America.[18] The following year he invited British climber Noel Odell—the same man who had drawn the distinction between glacial geology and glaciology—to join the team.[19] Odell was a geologist mostly known for his climbing skills and famous for being the last person to glimpse George Mallory and Andrew Irvine before they disappeared into the clouds on Chomolungma's (Everest's) upper slopes in 1924. Twenty-five years later Odell's climbing ambitions had not dimmed. With Snow Cornice, his research would focus on the base of Mount Vancouver and at the end of July he led three project members to its summit.[20] Project Snow Cornice, as initially envisioned by Wood, looked as much like a mountaineering expedition as it did a glaciological project. But Wood's vision would not be the only one to shape the project.

Walter Wood was a Swiss-educated child of an upper-crust New York family. He was a gentleman, a geographer, and a mountaineer, but he was no glaciologist. And he knew that. To run the project's glaciological program he enlisted Robert P. Sharp, a young California Institute of Technology geomorphologist with a penchant for ice. Wood met Sharp during the 1941 Wood Yukon Expedition and had worked with him during the war.[21] Sharp came from a markedly different background than Wood. The son of Southern California farmers, he was the first of his family to go to college. In 1930 his grades and gumption earned him a spot at Caltech, whence he went on to Harvard to earn a master's and a doctorate in geology. Unlike many of those involved in JIRP and Snow Cornice, who remained closely tied to Harvard, Sharp was a Caltech man through and through. He believed adamantly that the best training for earth scientists was in mathematics and physics and that basic research was valuable for its own sake, scribbling "Woe!" in the margins of a letter asking for the field's material and practical benefits.[22] "We study glaciers," he countered, "because we wish to understand them."[23] These, he maintained, were characteristics of a Caltech frame of mind.[24] Meticulous in his fieldwork and rigorous in his analysis and argumentation, Sharp forged a sterling scientific reputation. North American glaciology needed "more of that kind of thing," Field declared approvingly.[25]

Sharp had little interest in constructive mountaineering. Unlike Miller, Wood, and Washburn, Sharp drew clear boundaries between work and play. Which is not to say that he was no fun. Sharp loved fly-fishing. His grandfather had taught him how to do it and he enjoyed wiling away leisurely afternoons casting from riverbank or midstream. But, he sternly admonished one student, "you cannot go into the field thinking that at 5:00 p.m. you will lay down your geologic hammer and pick up your fly rod. A productive day must be fully devoted to one or the other."[26] A man of exacting standards for himself and others, when Sharp was in the field for research, he did research. He believed Wood's mountaineering attitude compromised Snow Cornice's scientific soberness. When invited to join the project, Sharp agreed but demanded autonomy over the glaciological aspects.

It wasn't just that Sharp objected to climbing on the job. He also harbored doubts about Wood's research plans. He knew that an Ahlmann-type regimen would be exceedingly difficult to fulfill. Good regimen work, as Ahlmann himself had stated, required a small glacier that could be studied from tip to toe. The Malaspina was a poor candidate. It is the largest piedmont glacier in the world. Piedmont glaciers flow unhindered onto wide flats, which gives them a distinct, clamshell-shape. At its widest point the Malaspina is today 65 kilometers broad and pours into the Gulf of Alaska like the oozing pseudopod of some enormous amoeba (fig. 14). Its magnitude and confluence with the similarly mammoth Seward Glacier were non-negligible obstacles to obtaining good accumulation and ablation measurements, and accounting for all outflows of melt water would be nigh on impossible for such a large glacier. Indeed, Sharp had criticized Ahlmann's work at Vatnajökull on these very points.[27] In general, Sharp was unenthused about regimen studies. He thought too little was known about too few glaciers and the factors that influenced their fluctuations to support tenable large-scale claims about climate (polar or otherwise) based on glacier regimens. He preferred close structural work that could achieve a high degree of precision. Studies of ice that examined how it flowed and moved in exacting physical detail were more to his liking.[28] Therefore, Sharp countered Wood's program of regimen studies with one based on problems of firnification (the process by which snow becomes firn and then glacier ice), englacial structures, and the mechanics of flow, the latter of which he would study by boring a hole into the glacier and measuring its deformation

FIGURE 14 The Malaspina's girth and piedmont shape are apparent from space. NASA, Goddard Space Flight Center, Greenbelt, Maryland, via Wikimedia.

using an inclinometer (an instrument used to measure angles).[29] "Personally," he wrote, "I am a little more interested in the structure of the glaciers themselves and in their mechanics."[30] The mechanics of ice held more appeal for him more than did its fluctuations, even when treated in Ahlmann's style.

Under Sharp, Snow Cornice's glaciological research was focused, sophisticated, and efficiently exercised. It included experiments with radar, audio-sounding, seismology, and crystallography.[31] These were the geophysical techniques required to study the internal structure and movements of a glacier. The data generated would not be represented in photographs. Rather, they would generate numbers that would then be graphed or arranged in charts. Or they would produce other kinds of visuals, such as crystallographic stereographs. In the publications of JIRP and in Sharp's work on the Malaspina, readers can find photomicrographs of névé and individual ice crystals, snowpack density graphs, charts for mass-balance calculations, and physical diagrams of all kinds. This change in visuals was partly due to the new geography of fieldwork. An icefield is a difficult subject to capture in a photo, especially one taken from the ground. It's just too big. But there was more to it than that. Photography is qualitative, and termini fluctuations are unhelpful and irrelevant for explaining either the overall present state of a glacier or the details of its component structures and movements. In geophysical glaciology, photography was demoted as a field practice because it did not meet new standards for evidence or answer questions of interest to working glaciologists.

But cameras did not wholly disappear from the icefields. The conditions under which these projects operated dictated that the practices of geophysical glaciology in Alaska had to be different than what was practiced by their prewar colleagues. Under these new conditions, photography was assigned a new role.

Paper and Planes

Both Sharp's research program and the one developed by Field and Miller for the Taku Glacier required researchers to stay on the ice for extended periods of time, probing multiple sites across the icefield. This marked a departure from the termini-focused geography of Field's photographic studies and those of his Alaskan glacialist predecessors. The Taku Glacier, although relatively accessible, was still a large glacier in Alaska and required time and money to reach, especially when coming from urban centers in the Lower 48. Both it and the Malaspina were exposed to severe weather and frequently pounded by storms. Researchers would need adequate shelter and escape routes if they were to succeed in their fieldwork. Meeting the scientific goals of JIRP and Snow Cornice under these conditions required considerable financial, material, and logistical support, and older solutions to these needs were quickly recognized as inadequate.

In 1948, during JIRP's first year of operation, funding came from a variety of public and private sources, including the American Alpine Club, the American Geographical Society, and Pan American World Airways—the latter was interested in assessing the Juneau's suitability for siting a summer ski resort. Local support came from the Taku Lodge, Gus George's Grocery in Juneau, and the US Forest Service. As with the economics of the Vauxes' glacier studies, there were instances of gift exchange. Miller and Latady befriended the O'Reillys, owners of Taku Lodge, who provided free accommodation in return for promotions-friendly photographs. Field and Miller also drew on their connections to mountaineering and exploration circles, garnering support from AINA, the HMC, the American Alpine Club, and Harvard's Institute of Geographical Exploration. A less hoity-toity backer, the Oscar Meyer Company, donated fifty pounds of canned meat. (The pork with barbecue sauce, beef with barbecue sauce, and wieners were declared "excellent," though this assessment may have been colored by the alternatives on offer: pemmican and "fish balls.") The Harmon Foundation, whose namesake, Byron Harmon, was a photographer of the Canadian Rockies, donated photographic supplies. Finally, each party member chipped in $125 of his own money. In terms of the material and financial support it received, JIRP's initial year was like the proverbial child: raised by a village, and a rather eclectic one at that.[32]

Anyone might have mistaken that first year of JIRP for a mountaineering enterprise. Its ragtag assortment of sponsors was cobbled together in the fashion of a climbing expedition, and almost all its crew were mountaineers. There was good reason for this. The high altitude glacial setting of the Juneau Icefield required knowledge of mountaineering techniques. Jirps, as they called themselves, needed to be able to travel "on rope" and rescue team members if they fell into crevasses. They had to know how to read ambiguous signs of mercurial mountain weather and how to cope with whatever blew in. Mountaineers are not the exclusive keepers of such wisdom, but since the early nineteenth century they have cultivated skills and knowledge that allow them to live and work in the high alpine. Folk like Mal Miller were well-suited to JIRP's geography. They also wanted to be there. Miller was ever on the spy for an enticing line or a clean pitch of rock; he believed a little "off the clock" climbing was good for group morale.[33] Walter Wood believed it was good when on the clock.

Field and his predecessors, though often trained in the ways of mountaineering, spent little time *on* tidewater glaciers whose termini were more easily reached by boat than by scrambling to high vantage points. Not so for JIRP and Snow Cornice scientists who, like the Vauxes, needed expertise in technical climbing. However, the Vauxes' expeditions were small-scale and required only supplemental financial backing to get a handful of glacialists to the ice and back. And their focus on repeat photography of glacier termini meant they spent only some of their time on the ice. JIRP and Snow Cornice were on a much larger scale in terms of geography, research ambitions, and personnel, and both projects took place almost entirely above the treeline. In this instance, quantity altered quality, making geophysical glaciology in Alaska something altogether new.

Turning the Taku's névé into an instrumentalized glacier demanded reliable funding. There was only so much reconnaissance work that Pan Am Airways would fund before deciding whether the Juneau Icefield offered good commercial skiing (turned out it didn't). JIRP also needed salaried staff. The project would not be run by professors with permanent positions and their own research commitments; it would need young scientists who could devote their summers to fieldwork under someone else's direction. These men couldn't be asked to do so for free, but the AGS couldn't afford to pay them. In 1948 expedition members had contributed some of their own money, but this was undesirable and unsustainable. Moreover, the research planned for 1949 embraced an even broader range of activities requiring more equipment and more personnel. To fulfill its aspirations, JIRP needed more, and more stable funding and a reliable way of getting everything and everyone to and from the ice.

The solution was military patronage. In October 1948, as staff were beginning to process the summer's field data, JIRP received $20,000 from the Office of Naval Research (ONR), the defense institution responsible for hydrological research. François Matthes would have been pleased. The initial contract lasted from May 1949 until April 1950, with the possibility of annual renewal, which it received every year until 1958. Robert Sharp, who had suggested to Mal Miller that JIRP solicit the ONR, looked to the same agency for fiscal independence to undergird his scientific independence from Wood's constructive mountaineering agenda. He had applied for funding from the ONR's Pasadena office earlier that year. Wood, who had garnered military assistance for his mountaineering expeditions even before the war, had already secured ONR support for Snow Cornice. Blocked at its rear by precipitous heights and at its terminus by deep and tumultuous waters, the Malaspina's location rendered the project entirely dependent on air support.[34] Geophysical glaciology in Alaska, where the mountains were daunting and glaciers inaccessible, required backers with money and aerial capabilities. The military had both.

From its inception in 1946 until 1950, when it began to be eclipsed by the National Science Foundation, the ONR was the principal supporter of scientific research in the United States.[35] Expeditionary science in extreme environs was an ONR priority, particularly for the geophysics branch.[36] Still, this doesn't answer why *glaciologists* received military benefaction. Why would an organization devoted to the high seas want to pay for studies of Alaska's ice?

The answer is rooted in the recognition of the strategic importance of ice and snow during and after the Second World War. Military interest in cold places continued after August 1945 because the burgeoning Cold War rendered high latitudes complicated geopolitical zones. In Antarctica a veneer of scientific cooperation barely masked nationalist rivalries and territorial jockeying, both rooted in the possibility of finding alcoves of mineral wealth buried beneath the ice.[37] In the Arctic tensions across the polar Iron Curtain provoked the United States and Canada to look northward. Anticipating Soviet attacks via the most direct line over the Arctic Circle, North American governments strove to transform arctic lands and waters into what historian Janet Martin-Nielsen terms "landscapes of prediction and control." To control the North geopolitically, one had to first know the North.[38] Field scientists, too, typically seek to impose controls on the spaces in which they operate.[39] But they work in places that are emphatically beyond their control, as exemplified by Ahlmann's experiences on Vatnajökull and by the antics of the Vauxes' movement plates. This shared search for epistemic control in unruly places makes the military and the field sciences cozy bedfellows. As historians have documented, government strategists, military hawks, and scientists worked to make the Arctic and sub-Arctic into zones of militarized knowledge production.[40] Glaciology, the science of snow and ice—the elements that define polar environments—was integral to this task. Like other geophysical and geographical sciences in the early Cold War, glaciology became geopolitics by other means.

We know quite a bit about what this looked like later in the Cold War. For instance, at Camp Century (1960–66), where the US Army bored tunnels into the Greenland Icecap, from whence they hoped to secretly monitor and counter Soviet activities. They installed, among other things, a portable nuclear reactor, only to learn that glacier ice will mercilessly crush human burrows and reactors alike in its unwavering plodding toward its margins.[41] Historians have also investigated glaciology's place in the International Geophysical Year (IGY) (from 1957 to 1958). The IGY was an eighteenth-month-long international effort to study the Earth—its oceans, tectonic plates, stratosphere, ionosphere, volcanic aspects, among other systems—using the tools of geophysics. It is often recognized as an important moment in histories of the earth and environmental sciences, as well as one of the most instructive instances of science operating as geopolitics in the twentieth century.[42] It is also often heralded as the beginning of modern Antarctic research, and histories that treat IGY's glaciological aspects often focus on the Antarctic program as the seed of modern glaciological research. But before Camp Century and before the IGY, there was the Juneau Icefield Research Project and Project Snow Cornice, which demonstrate that geophysical glaciology came to America earlier, on the very heels of war.

Their wartime service made glaciologists wise to the strategic importance of snow and ice, and they adroitly manipulated Cold War anxieties to their advantage. Rumors of a vigorous Soviet glaciology abounded. "The Russians are known to be making observations of the glaciers," wrote Field to potential funders. "Little is known of the results, but it seems likely that they have put more effort into systematic long-term programs . . . than has been the case in this hemisphere."[43] Outlines for research programs noted that American glaciology lagged behind the vigorous work being carried out in Europe, making certain to single out efforts in the USSR. To be fair, Soviet scientists *were* researching snow, ice, and permafrost.[44] But until the mid-1960s their American counterparts knew of little more than hearsay evidence and the translated titles of publications. Still, they employed this partial knowledge to their advantage. Closing the perceived glaciology gap would boost the national ego and serve the security interests of a nation for which the polar regions had become sites of potential conflict and theaters for actual scientific competition.

Glaciology received money because it promised to provide knowledge of how snow and ice worked in environs that the American military establishment perceived as important. But money was not the only gift the military gave glaciologists. It also gave them wings. Without aerial assistance, North American geophysical glaciology would not have gotten off the ground (pun intended) in Alaska, no matter how enticing the scientific puzzles posed by its glaciers.

Historian Vanessa Heggie has shown that, historically, altitude is generally more restricting for research than is the extreme cold of lower altitudes, such as those found in Antarctica.[45] Altitude alone can slow cognition and basic physiological processes; mountain topography makes travel difficult; and both of these effects are compounded by cold temperatures. Many challenges, however, are rendered less daunting by transportation that effectively moves people and supplies to and from mountains speedily. In the Canadian Rockies the Vauxes had the Canadian Pacific Railway; in Scandinavia, Ahlmann had dogsleds and ponies; in Switzerland, portals bored through glaciarized mountainsides allowed scientists to disembark midway through train tunnels hewn through mountains. Southeastern Alaska, home to North America's most puzzling ice streams, had none of these advantages. Long-term, on-site fieldwork there was made possible by military air support. Aerial support allowed the large quantities of supplies and equipment demanded by geophysical research to be transported to the icefields for prolonged stays. The next likely hauling option in the Alaskan back country was pack train or dogsled. But ferrying supplies for endeavors as large as JIRP or Snow Cornice would have left man and beast at the mercy of changeable weather for more prolonged periods, increasing risk of peril for both while cutting into valuable research time and prolonging the expeditions. Walter Wood's 1935 Wood Yukon Expedition required nine days to haul two weeks' supplies for four men to the base of Mount Steele. Following Bradford Washburn, who had used air support on a 1934 expedition to Mount Crillon, in 1941 Wood convinced the War Department to supply two Army Air Corps B-18-A bombers and one captain, Albert H. Jackman, to make parachute drops. In two days Jackman free-dropped and parachuted 1,656 kilograms of supplies at the mountain's foot.[46] Thus, even before the war, as a mountaineer Wood learned that aerial support made it possible to furnish larger expeditions with more materials in less time. Drawing on this experience and a canny grasp of what would be of interest to military backers honed during his wartime service, Wood turned the drawback of the Malaspina's inaccessibility into its virtue. He presented it as an opportunity for experimentation in logistics, in equipment performance, and in the construction and maintenance of airstrips on ice in remote, elevated environs.[47] These were topics of interest to a military establishment facing the prosecution of a cold war on icy, frozen lands.

The Taku Glacier was more accessible than the Malaspina—as noted, the icefield could be reached by foot and by boat. Nevertheless, aerial support was still critical for JIRP. Flying personnel to the Juneau Icefield was not strictly necessary, but the supplies needed for long-term geophysical research were too substantial to haul by dogsled or ski. During its inaugural season, before ONR patronage had been secured, the military provided limited aerial transport for personnel and supplies. On September 5, just before the onset of bad weather, a Neptune Bomber out of the naval air base at Ketchikan dropped over 680 kilograms of equipment on the névés of the Taku and Twin Glaciers.[48] This sounds like a lot, but it was small potatoes. The following year, the naval air station at Kodiak made fifty-five parachute drops and three hundred free-fall drops using PBY and R4D aircraft to place 12,000 kilograms of building materials on the ice. More fragile materials, including 300 pounds of seismic equipment, were set on the névé by a ski-wheeled C-47.

The construction figure alone dwarfs the previous season's measly total of 680 kilograms of gear. It suggests that in 1949 Jirps were busy populating the ice with built structures. And they were. The biggest building was a research hut erected "on the brow of a westerly protruding cleaver of rock."[49] The fourteen-by-twenty-foot hut at Camp 10 was framed with two-by-fours, insulated, and clad in corrugated aluminum. It provided a sometimes too-cozy sleeping space for researchers as well as shelter for books, instruments, and gear. During the long summer days the Camp 10 hut functioned as quarters for cooking, writing, and escaping inclement weather, which was as heinous as it had been in 1948. Without airdrops, Camp 10 would not have been built or would have taken much longer to build. Fewer instruments and supplies could have been housed and researchers would have been more exposed to changeable weather and unable to spend as long on the ice. Air supply enabled Jirps to stay on the Juneau for thirteen weeks that year, ultimately peppering it with thirteen more camps.

Aerial support made long-term in situ studies of remote Alaskan glaciers easier to supply, execute, and staff. Military support may not have been *essential* for geophysical research on glaciers—Richard Goldthwait's seismic work in 1935 is a good counter-example—but it was essential for geophysical research done on the scale of JIRP and Snow Cornice. Like many cold war earth scientists, the organizers of JIRP and Snow Cornice welcomed this flush of funding. They did not see military money as "tainted" or inhibiting, but rather as enabling. Certainly the ONR was a better patron of science than, say, the tobacco or fossil fuel industries. The navy was interested in obtaining accurate information. This is not always the case for all patrons of science, some of whom seek disinformation to support their products or views.[50] But because it wanted facts, Sharp perceived the ONR as a means toward research autonomy. Even Field, who could have complained about the ONR axing the Patagonian half of his glacier research project, instead focused on the opportunities proffered by navy money. ONR contracts saved him from having to cobble together funds from airlines or wiener companies or from asking young scientists to contribute their own funds. It is no wonder, then, that glaciologists welcomed military support. It was enabling and was experienced as a good thing.[51] I have found no evidence of them wishing things were otherwise. Of course, absence of evidence is not evidence of absence, but the available evidence suggests that many glaciologists' research goals dovetailed with military interests. The physical workings of snow and ice were priorities for both navy patrons and glaciologists, which is why Sharp could turn to the ONR when he wanted to protect the sanctity of his research: it was a confluence of interests based in a shared desire to precisely understand frozen landscapes.

Consequences of Military Patronage

To say the situation was a confluence of interests does not mean that military patronage did not impact glaciologists' practices or shape their agendas. Nothing is free. All patrons fund research for a reason, and these reasons cannot help but affect the research. Historians of science have demonstrated a variety of ways this occurs. There are well-known cases wherein military patronage did affect what was studied and what knowledge was produced. The building of the atomic bomb vividly demonstrates how military money can guide objectives for technological development and orient research questions and resources toward a particular goal. Military money can also shape research in ways that have downstream effects. Paul Edwards has shown how military objectives guided computing science down particular channels, shaping the subsequent trajectory of the nascent computer industry.[52] And Naomi Oreskes has shown how military patronage of American oceanography meant that certain questions were pursued—those having to do with the chemistry and physics of the oceans—while others—relating to biological oceanography—were not, or not as much. This has affected what is known and ultimately what could ever be known about the oceans, given the subsequent loss of marine biodiversity in the twentieth century.[53] In other cases, military patronage has helped shape how scientists and broader publics understand the goals and objects of study, including nature itself.[54] Military patronage, then, can have multilevel and varying effects on scientific research. The question for historians, Oreskes suggests, is not whether military money does or doesn't affect science, but rather, "How did this particular form of patronage affect this science at this time?"[55]

How, then, *did* military patronage affect geophysical glaciology in its early North American instantiations of JIRP and Snow Cornice? For American glaciologists, as it was for their oceanographer colleagues, military patronage was both enabling and constraining. It made it possible for them to do the type of geophysical work their European counterparts were doing in the remote and rugged conditions of coastal Alaska. But it also shaped both their practices on and off the ice and their research agendas, with consequences for subsequent glaciological knowledge. ONR patronage altered the intellectual property regimes and authorial practices of geophysical glaciology. The case of Mal Miller illustrates this best, as rules become most visible when they are broken.

As a stipulation of its JIRP contract, the AGS owed the ONR annual progress reports and copies of any materials produced by the research. This included all writing, photographs, films, maps—representations of all kinds. Any publication relating to the project first had to be cleared by the navy. Miller knew this because he and Field had together drafted the AGS's policy on publications, which mandated articles be submitted to the society prior to being sent to a publisher. Yet, like Washburn, Miller came to glaciology as a mountaineer and was inclined to treat the field as a site for adventure, as well as research. His proclivity for publishing "trip reports" in popular venues like the *Bulletin of the Harvard Mountaineering Club* was a common practice among mountaineers that did not always sit well with the ONR. Of particular embarrassment to Field and the AGS was the 1949 *Science Illustrated* article "Vanishing Glaciers," which portrayed glaciology as a romping adventure in the mountains. To Field's mortification, Miller included a classified photograph of an upturned Norseman plane that had somersaulted over its nose while delivering supplies. Miller believed such articles attracted public support—a technique commonly used by mountaineers seeking newspaper backing in the nineteenth and early-twentieth centuries. Field saw them instead as a violation of ONR and AGS policy.[56]

Military patronage did not utterly kill the practice of publishing in mountaineering venues, but it did constrain the relationship between field scientists and their data.[57] Contra what Miller would have liked to have been the case, they were not free to do whatever they wanted with "their" data. Times had changed from the days of glacier naturalism, when individuals like Reverend Green could write long-winded chronicles interweaving descriptions of mountaineering, scenery, and scientific results. Now, glaciologists had to clear their publications—scientific and recreational—with their military backers. They had secured financial stability at the cost of authorial autonomy. Young Mal Miller, apparently, balked at this new constraint.

Military funding also constrained what glaciologists could study and what would ultimately be known. As mentioned, when JIRP was still the transhemispheric Glacier Studies Project, the ONR chose not to fund the South American portion of the project. Patagonia was not as relevant as Alaska to American military interests. The AGS could not support such expensive field studies alone. The Patagonian portion was dropped, and the Glacier Studies Project became the Juneau Icefield Research Project. In the process, an opportunity was lost to produce the global comparative knowledge that Ahlmann had urged glaciologists to seek and that might have helped identify trends of global warming more quickly and with greater certainly. As with cold war oceanography, military input meant that scientists would learn certain facts, but not others. Early comparative knowledge about glacier behaviors in different hemispheres was lost because of what the navy deemed imperative.

Yet ONR priorities did more than just dictate what glaciologists would not learn. Awareness of military interests nudged glaciologists into learning things they otherwise would have not learned, modifying how they spent their time on the ice and how they used their cameras. The physics of snow and ice was not all that the ONR wished to know about. It wanted information on how to survive and operate in extreme, cold weather environments. Thus, military patronage encouraged glaciologists to prioritize questions about rations, instrument maintenance and performance, and transportation as important research objectives in the field.[58] When first applying for ONR money, Field emphasized the Glacier Study Project's value for training, equipment testing, and developing procedures for operations, logistics, and communications in frozen environs.[59] Walter Wood committed Snow Cornice to similar objectives, adding investigations of glacier surfaces for airstrip construction. Emphasizing the potential to produce information relevant to operations in frozen landscapes was an adroit tactic for courting military money. Yet it also committed glaciologists to broadening their research agendas to include questions that were not glaciological, geological, or meteorological in nature. Jirps made studies of equipment and rations for the Quartermaster Corps; Wood's crew reported on the efficacy of air supply to remote mountain stations. This labor was a form of what could be termed paraglaciological research. It certainly was orthogonal to questions about snow and ice. But it was a newly obligatory element in military-funded glaciology. While military patronage might not have significantly altered the *glaciological* questions scientists asked—for those questions largely met in a confluence with navy interests—it did encourage them to spend time and resources observing and documenting the organizational logistics and materials that enabled them to live and work on the ice. The legacy of this shift in research agenda and field practices was documented in the visual culture of glaciology, where a further consequence of military funding was a new role for photography.

From Termini to Tasks

In postwar glaciology, techniques and capabilities developed in tandem. Geophysical techniques like crystallography, seismology, and vertical borehole drilling enabled researchers to study the physics of glaciers; military patronage made it possible to get the required instruments to the ice and support their extensive use. Under these circumstances, the camera became one of many tools in the glaciologist's kit. But, unlike the seismograph or ram penetrometer (a device which measures the resistance of snow layers), photographs did not provide quantitative information. The camera was, thus, poorly suited to the needs of those interested in studying glacial structures and processes. This is suggested by the scarcity of repeat termini photographs in the publications of JIRP and Snow Cornice.

On the Taku and the Malaspina, the practice of glaciology was quite different from its prewar antecedents because scientists understood themselves to be investigating a different kind of thing than that of their predecessors. Geophysical techniques instrumentalized the glacier. The instrumentalized glacier was a different object of study than the glacier of glacier naturalism because researchers' practices and assumptions reshaped it as a place of inquiry. It was not approached as a fluctuating remnant of past ice ages that could be understood through repeat photographic documentation of its terminus. The glacier, as object of geophysical knowledge, was an aggregate collection of quantitative data that produced different kinds of evidence. Seismic measurements, inclinometer readings, and temperature-density charts constituted the evidence produced by JIRP and Snow Cornice. The instrumentalized glacier was treated as opaque to vision, even when augmented by photography. Cameras could not capture quantitative information so they were deemed less useful in the study of the ice itself. But glaciologists still had cameras on hand. They just used them for a different purpose.

Far removed from termini and their fluctuations, glaciologists applied their cameras to capturing the details of operations, techniques, and supply: the fieldwork of interest to their military patrons (figs. 15, 16, and 17). This was deliberate and strategic. In his application to the Research and Development Board, Field made clear the value of photography in geophysical glaciology: "Photographic records of the Ice Field Operations," he wrote, "may be useful for training manuals and films."[60] His wartime experience making films for the army gave Field a certain authority in this domain. Photographs could, of course, be used to illustrate public and academic lectures, but, as Miller's experiences with *Science Illustrated* demonstrate, they first had to be approved or they became problematic. Photographs of operations on the ice were aimed, first of all, at attracting and retaining the interest of military audiences.

The emphasis on men, techniques, and operations can be seen in still and moving films made during JIRP's decade of operations. Silent, unedited reels show young men in plaid button-ups and scruffy beards cavorting for the camera, eyeing it nervously whenever its glance lingers too long. In the jolted, overly quick movements of retro film, the men dig pits and insert metal tubes into compacted

FIGURE 15 Mal Miller sprinkling lead oxide around a snow stake on the Juneau névé,
1948. From the Juneau Icefield Research Project Collections, American Geographical
Society Library Archives, box 165, folder 57. Reproduced with permission from the
American Geographical Society Library, University of Wisconsin, Milwaukee, Libraries.

side walls of snow pits, apparently gathering samples and measuring densities;
they peer through the surveyor's theodolite, then probe the névé with slender
metal wands. In other reels they pack boxes and canvas duffel bags onto sleds and
jerkily haul them across the snow on wooden skis. Other film clips show them
eating crackers or raising a wall framed in two-by-fours. The films splice abruptly
between such mundane scenes of life on the ice and oblique aerial views of the
icefield and surrounding peaks. The icefield is a smooth, gently rolling desert of

FIGURE 16 A JIRP staff member surveying, 1948. From the Juneau Icefield Research Project Collections, American Geographical Society Library Archives, box 165, folder 57. Reproduced with permission from the American Geographical Society Library, University of Wisconsin, Milwaukee, Libraries.

FIGURE 17 JIRP researcher measuring the height of the névé on a graduated snow stake, 1948. From the Juneau Icefield Research Project Collections, American Geographical Society Library Archives, box 165, folder 57. Reproduced with permission from the American Geographical Society Library, University of Wisconsin, Milwaukee, Libraries.

white pierced by pinnacles of black rock. These scenes make it clear that a camera at ground level was unhelpful for capturing large, featureless expanses.[61]

In geophysical glaciology, photography never ceased to be part of the field scientist's tool kit, but it was deployed differently than before. Only one of many instruments, the camera did not produce the quantitative information that weigh scales, inclinometers, and seismographs did. Instead, it captured on film the work done with those instruments. The changes in research agendas and standards of evidence that came with the culture of precision advocated by the champions of geophysics and encouraged by military patronage altered the status of photographs as conveyors of scientific information about glaciers. Graphs, charts, and equations became the preferred representations for depicting what was now of interest in glacier study: fluctuations in overall volume and the processes by which snow and ice transformed and moved. Photographs played a merely documentary role, capturing not changes in ice but changes in method, field site, and practice. To be sure, scientists photographed ice. But little could be seen beyond the sweeping prairies of sparkling snow. Aerial photography, although good for reconnaissance and for producing maps, was initially understood to be of limited glaciological value.[62] The primary value of photography to JIRP and Snow Cornice was its ability to document new modes of research on terrain of military interest and to use this documentation to convince patrons to fund such endeavors. They portrayed Alaska's glaciers as strategic places for technoscientific instruments and men: places that could be tamed into disclosing quantitative, predictive information by technologies and American stick-to-itiveness.

Nevertheless, it wasn't all about pleasing the military. There were good scientific reasons behind this shift too. Termini fluctuations are not reliable indicators of the overall present state of a glacier—mass balance measurements are much better. Contra the Vauxes' hopes, compiling more photographs of termini would not yield the secrets of how glaciers flowed and this turn away from repeat glacier photography had downstream effects for the accessibility of glaciological knowledge.

Instrumentalized Glaciers, Exclusive Glaciers

Repeat glacier photographs seem obviously legible. Their meaning seems as apparent as before-and-after advertisements for hair recovery products. Before: a thinning pate; after: luscious locks. Before: more ice; after: less. On the other hand, not everyone knows how to read a seismographic print out or mass balance chart. And, if you don't know how, it is immediately apparent that you have no clue what you're looking at. Unlike with repeat photography, you don't assume you know what you're looking at or presume to fill in the gaps of meaning. Instrumentalizing the glacier allowed glaciologists to make more precise and more accurate measurements than did repeat photography, but it also made the glacier an object about which only a few were qualified to speak and downsized the target audience to whom they spoke. Geophysical glaciology transformed glaciers from natural historical objects—which amateurs like the Vauxes or William Field could collect, catalog on film, and share with broad-based audiences—into instrumentalized landscapes to be probed and measured. These glaciers and the representations that captured them were understandable only to trained initiates. Thus, changes in visual culture that were part and parcel of a multistrand history shaping the practice of American glaciology marked and contributed to the building of a disciplinary silo around the science of glaciers. This was yet another way in which military patronage impacted glacier study, albeit indirectly.

Military support revamped how glacier scientists used cameras, which in turn shaped the visions of and therefore the understanding of glaciers. The photographs that were produced by geophysical glaciology, though different in form and content from the photographs taken by glacier naturalists, also conveyed ideas about what glaciers were, who belonged on them, and the reasons they studied them. Where the photographs of glacier naturalism depicted glaciers as wilderness landscapes for the adventurous exploits of wealthy tourists and genteel naturalists, the images of midcentury geophysical glaciology suggested that glaciers were the province of an even more select few: expertly trained scientists and technical mountaineers. Historian Peder Roberts has written that Walter Wood, in his scientific and mountaineering pursuits, inscribed his own ambitions and ideas onto the mountains. This, Roberts contends, facilitated, even necessitated the erasure of the Indigenous people who lived there.[63] We can generalize his findings to JIRP and Snow Cornice. The rich, specific knowledge of glaciers possessed by Tlingit, Eyak, Dene, and other peoples are not found in the writings or photographs of glaciologists. Although they were present, Indigenous colleagues and assistants are structurally invisible and their knowledge is quite literally papered over by glaciologists' representations of instrumentalized glaciers and the men who worked on them. Colonial assumptions and effects are no less legacies of geophysical glaciology than for glacier naturalism.

Moving On

In 1953, JIRP moved to the nearby Lemon Creek Glacier, just a ridge over from the Taku. The move was suggested by Ambassador Ahlmann, now a Swedish diplomat, who visited the Juneau Icefield in 1952 as part of an American tour that was both diplomatic and glaciological.[64] The Taku, it turned out, was not the ideal glacier, according to Ahlmann's criteria. The geometrically simpler Lemon Creek fit the bill much better. At the new site, new field leaders took over from Miller, the size of field crews dropped, annual budgets decreased, and research agendas narrowed. Jirps focused on regimen studies, surface movements, and the three-dimensional shape of the glacier.[65] Thus streamlined, the project continued under ONR contract until 1958. Overall, the navy contributed $202,651.50.[66] JIRP's final year coincided with the International Geophysical Year.[67] It was its smallest season to date; operations lasted only sixteen days, during nine of which the team was pummeled by driving rain and high winds.[68] Some things never change, including the good chance for nasty weather on an Alaskan glacier.

Snow Cornice never achieved Walter Wood's hope of becoming a permanent research station. In 1951 tragedy came to the Malaspina when the project's Norseman plane—the same one that had caused such embarrassment when it flipped on its nose—went missing. On board were veteran pilot Maurice King, Wood's wife, Foresta, and their daughter, Valerie. At the time, Wood and his son, Peter, were stranded on nearby Mount Hubbard, where they had been making another first ascent. They did not discover the plane was missing for several days. Searches yielded no sign of the Norseman and details of its passengers' fates remain a mystery.[69] That same year, Sharp began to lose interest in the Malaspina. He reported having "his stomach full of Yakutat" and being weary of the logistical challenges of doing science in such a remote location.[70] He began shopping around for a smaller, more accessible glacier that would better lend itself to three-dimensional analyses of flow. He landed upon the much smaller and simpler Saskatchewan Glacier, the same one that Field had photographed as a tourist in 1922.[71] Soon thereafter, as part of the IGY, he turned his attention to the Blue Glacier, where he focused on internal deformation and surface movements. The Saskatchewan, a small valley glacier that could now be reached via automobile road, and the Blue, accessed via a hike through the Hoh rain forest, were more manageable research objects. Sharp had small accessible ice streams that could be covered from tip to toe.

JIRP and Snow Cornice were the first large-scale sustained efforts at geophysical glaciology in North America. Their beginnings resembled interwar mountain-eering expeditions, but it was not long until researchers realized that geophysical glaciology in Alaska would have to be pursued differently. While they never lost their connections to climbing circles and practices, as the JIRP and Snow Cornice teams got deeper into geophysical glaciology, they evolved unique modes of fieldwork. The time and material needed for geophysical work, combined with the difficulties posed by Alaskan glacier névés and the state of postwar American science funding, meant that early geophysical glaciology on this continent would be supported by military patronage. ONR patronage impacted how glaciologists related to their data, how they spent their time in the field, and how they deployed their cameras. While photographs of glaciers were not terribly informative for investigating the physical properties of a gargantuan icefield, they were able to convey information about how to conduct the work and how to live on that icefield. Photography, thus deprived of its glaciological value, retained a documentary value in the context of military patronage. The products portrayed glaciers as exclusive spaces for technical, state-sponsored science. Although repeat photography was demoted as a form of evidence, the camera still functioned as a means of perpetuating certain ideas about glaciers, about Alaska, and about nature. Although the camera could not capture the instrumentalized glacier of military-supported geoscience, it continued to portray glaciers as empty spaces for science and mountaineering, inscribing certain kinds of knowledge and ends onto the land while erasing others.

4

MONITORING / *Environmental Glaciology*

On the evening of October 28, 1957, Richard Hubley, a rising star within American glaciology, stepped out of the Jamesway hut where he and his three companions, John Sater, Charles Keeler, and Robert Mason, had begun their overwinter stay on the McCall Glacier. High in Alaska's Brooks Range, the McCall is a long valley glacier hemmed by dark slopes that funnel ice north, toward the Beaufort Sea. Serviced by airdrops and Cessna landings from Fairbanks's Ladd Air Base, the group's work was the beginning of the McCall's life as the most studied arctic glacier on the continent. Their research station was perched in one of three basins that feed the main stream of ice. It was a harsh environment, scarred by lashing winds and the vicissitudes of high mountain weather. Yet it was hoped the four-man team would remain through winter. By late October, long polar nights were draping the McCall in cold shadows. In the darkness, 200 yards from the circle of light cast by the hut, Hubley, wearing only the bottom half of an arctic suit, lay down under the inky northern sky and allowed the cold to claim his life.[1]

Hubley's death was a blow to the study of glaciers in North America. Walter Wood wrote: "Dr. Hubley's death has come at a time when he has assumed national—and even international—leadership in his chosen field. North American science has been slow in building a coterie of scientists in the field of glaciology. Of those we have, Richard Hubley was one of the most distinguished. Science can as ill afford his loss as can we who knew him as a companion in the office and among the high snows."[2]

Few of his colleagues had known of Hubley's inner turmoil.[3] The aftermath of his death was marked by grief and confusion. Born in the city of Tacoma, Washington, alongside the flanks of its mountain namesake (Mount Rainier), Hubley came to first know glaciers in the peaks of his home state. He didn't stray far for his professional training, obtaining all three of his postsecondary degrees from the University of Washington. There, under the direction of meteorologist Phil Church, the Department of Meteorology and Climate (later the Department of Atmospheric Sciences) became a center for arctic climatology and glaciology in the 1950s and 1960s. Hubley was its second doctoral student.[4] At the time of his death he was poised to bring together the careful quantitative approach of geophysical glaciology and a concern for regional trends and glacier-climate research that had marked glacier naturalism. A new synthesis from old antitheses.

Hubley's death deprived glaciology of one of its brightest minds. Yet, through his life and then his death, he contributed to a sea change in American glaciology taking place in the years following the International Geophysical Year of 1957–58. He voiced an influential critique of the close studies of a single glacier epitomized by JIRP and Snow Cornice. These were deficient, he asserted, lacking comparative knowledge of many glaciers. How does one know whether the results obtained for one glacier can be applied to others? Glaciologists needed regional studies of multiple glaciers to place the results from individual ones in perspective. During his short impactful career Hubley tried to show that glacier-climate studies could be done with the quantitative precision of geophysical glaciology. Within this context, photography gained new value as a tool of regional monitoring. But now photographs would come from a different perspective: high in the air. As an homage to his deceased friend, Austin Post began making regular surveys of North American glaciers using aerial photography. His efforts would produce over one hundred thousand glacier photographs between 1958 and 1982. It was a fitting tribute. In 1953 Hubley had hiked out to get antibiotics when Post contracted pneumonia while on the Juneau Icefield, and Post had credited him with saving his life. A life's work for a life saved.[5]

Hubley's and Post's stories are a distillation of a larger shift that began in American glaciology in the late 1950s. The larger story revolves around a clutch of glaciologists that includes these two men: the Northwestern Glaciologists, as they called themselves, who coalesced around the University of Washington and the

USGS Glaciological Project Office in Tacoma. Even today an annual Northwestern Glaciology conference draws scientists from California to Alaska. After the IGY, these northwestern critics pushed back against what they deemed the myopic focus of geophysical glaciology as it was practiced on the Juneau Icefield and Malaspina Glacier in the 1940s and 1950s. They advocated, instead, for a combination of close physical work and broad-scale regional monitoring. Their research agenda was rooted in the precise, quantitative work advocated by Robert Sharp and other geophysical glaciologists but also made room for the bigger picture, which couldn't necessarily meet the earlier standards. After being sidelined in the decade after the war, photography reaffirmed its place in glaciological research. But now it would be situated in a new research context. Hubley aspired to produce regional aerial photographic studies that would contextualize individual glacier work. Post went further, putting aerial photography in the service of understanding the physics of glacier dynamics and demonstrating that although there were limits on what information photographs could generate, the antithesis between qualitative photography and geophysical glaciology was not set in stone.

The synthesis of geophysical techniques and regional monitoring produced a new regime of glacier research: environmental glaciology. My use of "environmental" to modify "glaciology" refers to a particular way of thinking about nature rather than an ethical commitment or political stance. It references a specific understanding of the environment as a planetary set of material and energetic relations among humans and the nonhuman world. Conceiving of nature as environment means seeing it as a dynamic web of systems and processes that link living and nonliving elements of every scale, including humans. Environmental historians have traced the roots of the idea to various origins, but these histories converge on the emergence of something novel in the twentieth century: a "conceptual revolution" that produced a way of seeing the world as unified biotic-abiotic whole, not just a backdrop for life but an indispensable, constitutive multiscalar framework.[6] They have shown how the idea of the global environment arose during the postwar era, out of cold war science and concerns about overpopulation, famine, nuclear winter, and chemical contamination.[7] Worries about population and resources led to the idea that the environment is closely aligned with resource monitoring and management. In environmental glaciology, then, scientists reconceptualized mountain glaciers as elements in the global environment; that is, as hydrological resources to be monitored.

Environmental glaciology arose alongside scientists' realization that humans were modifying the planet's climate. We might expect that in such a context, aerial photographs would be deployed as repeat images to serve as compelling evidence of global warming. But it was not so. It was instead another twist in the fortunes of repeat glacier photography. Glaciologists were busy using aerial photography to plan research projects and to scrutinize glacier dynamics; it was not the evidence that convinced them that global warming was real. Nor did repeat photography spur glaciologists' entry into scientific discussions of global warming. Glaciers' contributions to sea level rise was more of an impetus. Some of the most important early studies were made by northwestern glaciologist Mark Meier. Understanding glaciers as part of the hydrological cycle and intimately connected to the oceans—that is, as environmental objects—brought glaciologists into conversations about global warming, and their ideas were based in numbers.

The story of repeat glacier photography during this period in North American glaciology is, thus, surprising for us today, who have trouble seeing photographs of receding glaciers as anything but icons of global warming. Their fate during the years in which global warming was solidifying into a scientific fact illustrates once more how much historical context matters for how people see photographs. Like the glacier naturalists who saw the demise of ice age giants, the environmentally inclined glaciologists of the 1960s and 1970s saw something other than what we see today. The photographs do not speak for themselves.

Northwest Rising

Hubley's research on the McCall Glacier and the Blue Glacier in Washington state, where he had done his doctoral work, was a reply to and in many ways an outgrowth of Ahlmann's regimen work on circumpolar glaciers. Yet Hubley aimed to do better than the Scandinavian scientist. He shared Bob Sharp's concern that the data supporting claims about climate-glacier interactions weren't precise enough. He wanted an exact understanding of how ice mass and thermal energy were exchanged at a glacier's surface. Careful micrometeorological work—examinations of the micro-climates at a glacier's surface—combined with regional aerial photographic surveys, he believed, could address this problem. Aerial photographs would be able to tell you whether glaciers in the same neighborhood were behaving similarly in terms of gross movement and behavior. He brought this perspective to JIRP during the 1953 and 1954 seasons, and in 1957 he turned his attention to the Arctic. Careful and disciplined, Hubley was cut from the same cloth as Sharp. However, where Sharp homed in on the physical attributes of individual glaciers, Hubley did not lose himself in studies of ice crystals or the moisture content of a single square foot of snowpack. He appreciated the individuality of glaciers but studying one in detail would yield information on the factors influencing that glacier only. Scientists would be wrong to assume that one glacier could simply stand for others.

This problem is commonly attributed to the field sciences. As we have seen with glaciers in Iceland and Alaska, field sites are unique, sometimes unmanageable places. Their individuality and unruliness make it difficult for scientists to assume that knowledge generated in one place can be exportable to another.[8] This is not simply a problem for the field sciences. Science studies scholars have shown that lab work is never perfectly reproducible (it takes a lot of work to reproduce someone else's results), and that, moreover, fieldwork has exportable, or generalizable, elements.[9] All scientific knowledge is produced in specific places and must find ways to overcome the challenge of application beyond the conditions under which it was made. It often does, though not always in the same ways or everywhere.[10] In glaciology, Hubley and his colleagues proposed dealing with the problem of generalizing from knowledge obtained from individual glaciers by complementing it with studies of glaciers as a "species."[11] Like living species, glaciers may be grouped into regional "populations." Buffeted by the same environmental factors, individuals could be compared to others nearby to surmise whether the explanations for the behavior of one is likely to apply to its neighbors. Here, qualitative data, particularly photographs, could play a role. Thus, while he agreed with Sharp's critiques, Hubley did not take a "dim view" of observational studies as so much (mere) data collection with no useful result.[12] Repeat aerial photography could provide a backdrop against which to interpret the data from micrometeorological work. Indeed, Hubley maintained, if a micrometeorological study were to speak beyond the surface plane of any single glacier, it would need such a backdrop. He began with the glaciers of his home state.

In the late 1940s, USGS scientists determined that certain glaciers in the Cascades Range were gaining mass. Given that most known North American glaciers appeared to be receding, it was striking that those in the Cascades were growing.

Hubley took it upon himself in 1955 to find out if the trend observed by USGS scientists held more broadly. This was a serious undertaking. Outside of the ice on the stratovolcanoes Tacoma (Rainier), Wy'east (Hood), and Koma Kulshan (Baker), little was known about the mountain glaciers of the Cascades. There, mountain slopes are blanketed with thick forests from their summits to their river valleys which are hemmed by sheer walls, making travel overland tricky and arduous. Aerial photography proved a useful tool in such a place. Hubley took a series of aerial photographs over the North Cascades to compare with a limited set taken by USGS scientists five years earlier. The photographs revealed that fifty of the seventy-three observed glaciers were advancing. With a good deal of caution, he suggested this trend may be correlated with an observed increase in temperatures and precipitation in the Pacific Northwest over the previous ten years.[13] Seeing the big picture as well as the small, he made a case for the value of regional repeat photographic studies to complement the close work done on a glacier's surface.

With his micrometerological work and aerial photography, Hubley was poised to unify the research agendas of Ahlmann and Field, both of whom sought understanding of glacier responses to climate, and glaciologists like Sharp, who demanded precision, comprehension, and microscopic attention to ice. In so doing he helped push North American glaciology beyond research that was occupied solely with the structures and dynamics of ice. The "precise quantitative data" from a few glaciers needed "the less accurate information on many glaciers."[14] His work is celebrated for rekindling an interest in glacier-climate studies in North America.[15] Soon glaciologists on both coasts were calling for a combination of intensive studies and reconnaissance work.[16] Research on regional trends, one AGS employee predicted in 1960, was "destined to play a more important role in the U.S. glaciological program than [it has] in the past few decades."[17] Hubley should also be credited with sparking a renewed appreciation of the value of repeat glacier photography. The aerial photography program he began in 1955 was picked up and expanded by Austin Post, who took it to the next level.

Post was glaciology's most successful high school dropout (fig. 18). In the second half of the twentieth century he became *the* connoisseur of North American glaciers. He possessed an almost preternatural ability to recognize a glacier from its photograph and was able to identify virtually any ice stream from its portrait and to provide a date that was usually within two years. According to University of Colorado glaciologist Tad Pfeffer, Post's "understanding of glaciers was [. . .] intrinsic to his nature [. . .] unfettered by conventional training."[18] The consummate mountain man, Post preferred seclusion and the company of close friends. When awarded an honorary doctorate by the University of Alaska, the university's chancellor had to fly to Post's home on Vashon Island because the solitary glaciologist refused to attend the ceremony.[19]

Post took a circuitous route to glacier study. Born in 1922 to apple farmers in Chelan, Washington, he left school at fifteen to build trails and man fire lookouts for the US Forest Service (USFS) in the Chelan-Sawtooth Range. He spent the war as a navy carpenter's mate on Pacific campaigns. After being discharged, he returned to the USFS, and 1947 found him living the northwestern forestry life in the Upper Stehekin Valley mere miles south of Ross Lake, which Jack Kerouac would soon make famous in his literary celebration of that very lifestyle. One day a man in horn-rimmed glasses wandered into camp, looking for a climbing partner. Lawrence "Larry" Nielsen was a recent Cornell PhD who had grown up out west. He had returned to the region with East Coast friends, who turned out to be not all that interested in climbing. Post, an able climber and keen for adventure, volunteered to join Nielsen on nearby Mount Agnes. Afterward they stayed in touch, planning more ambitious expeditions to Alaska. Nielsen was a rheologist—a chemist who specializes in how materials flow. His day job was working in plastics for Monsanto, but his hobby was studying the flow properties of glacier ice, something that allowed him to combine his scientific expertise with his love of climbing. In a few years he would be Austin's ticket to Alaska's glaciers.

Calling Post a quick learner would be an understatement. His friends remember him as gregarious and formidable. "Always fun to be around," recalled colleague Birdie Krimmel, "his brain was always churning."[20] Trail maintenance, a monotonous if grueling job, could not hold the attention of such a mind. In 1949 he signed up to serve on the US Coast and Geodetic Survey's goodship *Derickson*. Displaying a proclivity for on-the-job learning that served him well his whole life, he spent the summer as the ship's carpenter, assisting with surveys of Prince William Sound. After returning to Seattle, Post spent the next few years as a carpenter. Then, in 1953 Nielsen became field leader of JIRP, and he asked his old Chelan climbing partner to join him in Alaska. It was JIRP's first year on the Lemon Creek Glacier and the new location had to be surveyed and camps

FIGURE 18 Austin Post on his way to the Worthington Glacier, 1957.
Courtesy of C. Suzanne Brown.

set up afresh. Starting anew meant that he and the other research assistants had to be jacks of all trades. Post's surveying experience, as well as his carpentry and trail-building skills, came in handy. He proved especially helpful when it came to constructing Base Camp, which required cutting trail and hauling heavy packs up steep pitches of ridge and glacier.[21]

JIRP introduced Post to scientific glaciology and some of its leading figures. In addition to Hubley—after whom he would eventually name his firstborn—he met Bill Field, who became an important mentor. "It is entirely due to Bill Field's enthusiasm and support," he once said, "that the aerial photo work began and my interest in the glaciers and glacier photography evolved from an amateur's hobby into a serious science."[22] After turning his own amateur hobby into serious science, Field was well poised to help others do the same. In many ways Post's trajectory echoed Field's. Both were amateurs in a discipline that procured the trappings of a professional geoscience over the course of their careers. They stayed in touch for all of their long lives, and Field remained a helpful collaborator and source of information for his younger colleague, more than once spurring his mentee into publishing. As Tad Pfeffer remembers, Post focused on "discovery over dissemination" and found glacier dynamics so intuitive that "detailed exposition and analysis may have seemed, if not difficult, at least superfluous."[23] Post may have taken awhile to come to glacier study but, as with his earlier endeavors, he proved a swift learner.

A Program of Aerial Photography

Austin Post began studying glaciers in the lead-up to the International Geophysical Year (1957–58). Dubbed "the assault on the unknown" by *New York Times* journalist Walter Sullivan, the IGY is regarded by many historians as a pivotal moment in the development of the earth sciences. It might well be assumed that this was when glaciology became geophysical. Yet, JIRP and Snow Cornice demonstrate that a geophysical approach among North American glaciologists had earlier, more diffuse origins. By 1951, six years before the IGY, North American ice streams had been subjected to seismic blasts, radar pulses, vertical boreholes, and crystallographic studies. Well before the unknown was assaulted, the champions of geophysical glaciology were already at work in the field. The IGY refined and amplified such efforts but did not initiate them.

The IGY did, however, break ground in the Southern Hemisphere, where it constituted the first sustained glaciological investigation of Antarctica. For this reason the IGY seemed to be such a revolution. Antarctica was indeed a geophysical unknown, and it dominated the attention of contemporary media as well as many subsequent historians.[24] Yet glaciologists working in the Northern Hemisphere saw the "much ballyhooed" escapades of their antipodal colleagues as a waste of money, personnel, and resources.[25] Walter Wood complained that it was profligate to devote 75 percent of American glaciology's meager population to antarctic problems when more useful work could be accomplished through logistically simpler studies nearer to home. This is an ironic comment coming from the man who had organized Snow Cornice on the inaccessible and logistically challenging Malaspina Glacier. Still, he was right that antarctic research was expensive and lacked the foundational knowledge that had already been established for northern glaciers.[26] For glaciologists not working in Antarctica, the IGY was less a foray into the unknown than an extended application of tried-and-true agendas and methods.[27]

These old methods now in the hands of young scientists sparked a reappraisal. Hubley's calls for qualitative studies on regional scales did not fall on deaf ears. During and after the IGY there was a sense among American glaciologists, particularly those of the northwestern school, that they needed to attend to regional trends to make sense of results from individual glaciers. Field, who had continued to make photographic studies even as he learned the value of close, physical work, was keen to support qualitative, comparative work from his post at the AGS. In January of 1959 he proposed a "program of aerial photography of Alaskan and western U.S. glaciers" to Phil Church. Church passed it to Post, who appreciated how important it had been to his late good friend Hubley; he had been looking for such a project. But first he needed to secure funding. The Department of Meteorology and Climatology's budget was at capacity with other projects.[28] Post did not realize this, but it was an ominous portent for later years. While the ONR had been the predominant source of money for geoscience in the aftermath of the war, the National Science Foundation (NSF), created in 1950, soon eclipsed other patrons in funding American science. After letting Post struggle on his own for awhile, Church eventually seized the reins, proclaiming, "Here let me handle this! I'm sure the NSF will approve the proposal if WE [University of Washington] submit it!"[29]

He was right. After receiving approval on August 5, Post leapt into action, enlisting Port Angeles–based pilot William Fairchild to take him over Washington's Olympic Mountains to try out his Eastman Kodak K-24 aerial camera and equipment. The K-24 was a wartime technology developed for night and aerial reconnaissance photography. It was a more lightweight version of earlier British-designed aerial cameras, but still weighed almost fourteen kilos and required two hands to use. By August 15 Post was flying to Fairbanks with his K-24 and an additional forty-one kilograms of camera equipment. From there he made his way to his most important objective: Henteel No' Loo' (the Muldrow Glacier), a broad ice highway that winds down Mount Denali's eastern and northern slopes. In 1959 it was of interest because a substantial surge had occurred during the winter of 1956–57, sending the glacier birling down its bed in the span of a few brief months.

About 1 percent of the world's glaciers surge, which means ice flows rapidly toward the terminus, occasionally up to one hundred times faster than usual speeds. Surges are often betrayed by an increase in crevasses, which form under the stresses and strains of rapid motion, like ripples in moving water. When IGY scientists noticed a dramatic increase in crevasses on the Muldrow, Post and others, including the US Navy and Bradford Washburn, who were nearby for other IGY projects, rushed to capture the event with their cameras. Post's documentation was deemed "the most complete." In 1959 he was returning to follow up on that work. Supporters on both coasts wanted to see him be the first to publish. "Do everything possible to assist, expedite or even prod Austin to get his data into print," one colleague urged Field.[30] Already, Post's reputation as a gifted observer and interpreter but reluctant author and self-promotor was recognized. His analysis, published in the *Journal of Geophysical Researches,* compared the Muldrow to similar events on the Black Rapids and Sustina Glaciers. He suggested that surges were not the result of weather anomalies or earthquakes (as earlier explanations had stated), but rather were an expression of the internal dynamics of glaciers that had certain characteristics.[31] Post thus went beyond Richard Hubley's hopes for glacier photography, using repeat aerial photography for more than just regional comparative studies. He put it to work in understanding glacier dynamics.

With renewed NSF funding, Post took some four thousand aerial images over the next three years, often while dangling precipitously out the side of Fairchild's aircraft. Like with his work on Henteel No' Loo', he was able to make his photographs do more than simply record changes in the glaciers' shape and size. He scrutinized them for clues about dynamics. His ability to do so and his uncanny ability to recognize glaciers from photographs were winning him a standing among glaciologists. However, in science, reputation does not always translate into reliable funding. Throughout the 1960s Post spilled much ink seeking funding for his glacier photography. In 1964 the NSF did not renew his grant. Although North American glaciologists appreciated the need for such work, it seemed that no one wanted to pay for it. A preference for proposals that produced quantitative data still reigned among midcentury funders of the earth sciences.

Financial stability for Post's photographic work eventually came from tectonic instability and from a man named Mark Meier. On Good Friday 1964, North America's largest recorded earthquake shook Prince William Sound and the Chugach Mountains, liquifying soils and plunging coastal forests into the salt water below. The 9.2 quake made the Seattle Space Needle quiver and was registered on seismographs in every state but Connecticut. Amid all of this upheaval and destruction, Post's northwestern school colleague, Mark Meier, sensed opportunity.

Meier was a student of Bob Sharp and a key player in northwestern glaciology (fig. 19). An Iowa farm boy characterized as "quiet, with a dry sense of humor, not shy, but not a pusher," Meier had "won over" the Caltech geology faculty on the merit of his master's thesis at a time when the university rarely accepted graduate students not already courted by professors.[32] Sharp trained Meier in the Caltech style, prioritizing the study of physics and mathematics and instilling an exacting approach to scientific problems and pronouncements. His dissertation, "Mode of Flow of the Saskatchewan Glacier, Alberta, Canada," was a distillation of geophysical glaciology's ideals. Sharp approvingly declared it the most comprehensive physical study of a North American glacier to date.[33] In 1960 Meier was invited by Luna Leopold, head of the Water Resources Division of the USGS, to head the first USGS glaciology office. The USGS had long supported glacier research, beginning with Israel C. Russel's studies in Alaska, but none of the early work was organized under an umbrella of glaciology. The Tacoma office was established in 1956 as a step in that direction; Meier was its first staff member.[34] Wendell Tangborn, who worked there with Meier for many years, recalled his employer's drive and ambition, traits, Tangborn recalled, that could make his boss occasionally

FIGURE 19 Mark Meier surveying the South Cascade Glacier, circa 1962. Photo by Wendell Tangborn. From the collections of W. Tad Pfeffer.

difficult to work with. Others remembered a friendly but removed mentor who seemed to prefer to work alone.[35]

At the USGS, Meier entered a job where the outlines of his research program were already sketched for him. This thoroughly Caltech scientist, trained by a man who believed that scientists study glaciers simply because "[they] wish to understand them," had to reorient his training in glacier dynamics to instead monitoring glaciers as hydrological entities. The USGS was not interested in glaciers as the source of elegant physical problems but rather as elements in the water cycle. Water resources had been a central concern for the Water Resources Division since its founding.[36] In 1960 Leopold had remained true to this vision. He wanted to know more about how glaciers connected to other elements of the hydrological cycle, such as precipitation rates and watersheds, and how they were affected by climate. Research at the new glaciological office was organized around three questions: How do climatic factors affect the regimen of a glacier? How do glaciers respond to changes in their regimen? How can glacier fluctuations be interpreted on the basis of climatic change, and vice versa?[37] Meticulous studies of the microscopic shifts in crystalline lattices of ice under pressure were not unwelcome, but they needed to refer back to the basic problem of how glaciers gained and lost mass in particular climatic conditions.

Meier brought a Caltech physicist's sensibilities to his new situation. As he saw it, the problem of how glaciers interacted with climate could not be divorced from questions about how they moved, gained and lost mass, and were theorized as dynamic objects. His first major project at the USGS was a long-term hydrological and mass balance study of the South Cascade Glacier, northwest of Seattle. Initially undertaken as part of the IGY, Meier hoped to use the South Cascade project how scientists did mass balance studies. It was a "gross simplification," he maintained, to speak of the growth or diminishment of a glacier as a whole.[38] Was the mass balance of the South Cascade determined more by winter accumulation or by summer loss? Did certain parts of the glacier respond differently to these factors? Answers to these questions would lead to more refined understandings of glacier-climate relations. The South Cascade was a good choice for this kind of work. It is a small, simple glacier that sits on impermeable bedrock, with only one outlet stream. Thus, unlike Vatnajökull or the Malaspina, all its melt water can be accounted for. The South Cascade project became a crown jewel in Meier's professional legacy, as well as in that of the USGS Tacoma office. Today it is the longest continuous mass balance record of any glacier in the world.[39]

As the success of the South Cascade project demonstrates, Meier had a nose for good research. He knew that Post had photographed the glaciers of Prince William Sound in 1963. When the Good Friday earthquake hit the following year, he saw an opportunity to evaluate a long-standing hypothesis about why glaciers surged: the earthquake advance theory formulated by Ralph Tarr. After the 1899 Yakutat earthquake, Tarr had stipulated that such an earthquake would cause an avalanche to load snow and ice on the upper regions of a glacier, causing a delayed flood of ice to sweep down the glacier. No one had yet proved or disproved this theory, though Post had suggested another explanation based on his study of the

Muldrow. If earthquakes did cause glaciers to surge, the Good Friday quake, with an epicenter so near to Prince William Sound, would have affected many glaciers and Post would have had the trained eye to spot the changes. In 1964 Meier arranged for USGS funds to support an aerial photographic flight over Prince William Sound. The following year he hired Post as a full-time technician. The resulting analysis, published in *Science*, severely discredited the earthquake advance theory.[40]

Thus Post finally found professional stability at the USGS. He stayed on until 1982, using aerial photography to make annual censuses of glaciers of the North American West and North. To this end he modified Fairchild's twin-engine Beech 18 so that five twenty-nine-kilogram K-17 cameras could be mounted on the plane. These heftier cousins of the K-24 were also used during the war and were considered the standard for aerial mapping long after. Unable to afford supplemental oxygen, photographer and pilot were responsible for keeping each other awake in the freezing, low-oxygen environment at 5,486 m, a mere 700 m lower than Denali's summit. When the plane started drifting, Post would shout "Wake up, Bill!" as he leaned out the open doors, wrestling with one of his cameras.[41] It was a matter of making do and "McGivering"—something the self-taught polymath Post excelled at.

All told, between 1960 and 1993 Post produced more than one hundred thousand negatives of vertical (straight down) and oblique (above and to the side) aerial photographs of glaciers. The oblique photographs are chiaroscuros of black mountain peaks wrapped in gleaming white glacier scarves, capturing crawling, furrowed ice tongues and snowfields thick and soft-seeming as winter coverlets. Devoid of people, they harken back to the wilderness conventions of glacier naturalism, but from a god's-eye perspective. They deploy what art historians call "the magisterial gaze": an elevated perspective that leads the eye into epic landscapes suggesting even greater marvels beyond the frame.[42] The vertical photographs, on the other hand, are abstract, flattened perspectives (fig. 20). From their straight above vantage point it is difficult to distinguish glacier from mountain in the creased and dimpled textures; colors and lines don't match up. Mottled snow patches and snaking waterways appear like the fatty deposits and veins of some enormous dissection. No wonder it required the trained eye of a person like Post to make sense of these images.

Along with Field's, Post's is one of the largest collections of glacier photographs

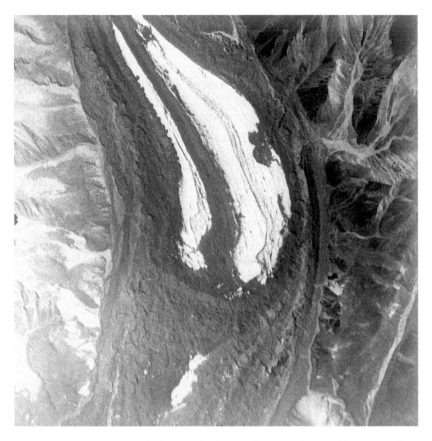

FIGURE 20 Vertical aerial photograph of Henteel No' Loo' (the Muldrow Glacier),
1970s. Photo by Austin Post. Glacier Photograph Collection, National Snow
and Ice Data Center, Boulder, Colorado.

on the continent. His photographs documented changes in glacier dynamics such
as surges and captured structural features such as snow lines, patterns of snow
accumulation, seracs, crevasses, ogives, eskers, and more. Often they bore witness
to recession. But this was not what Post used them for. Today, Post's photographs
circulate widely, often with re-creations produced by photographer John Scurlock.
It is hard to look at the recession captured in Post's work and not think of global
warming. The photos seem like prescient visual eulogies in a world mourning
glacier "deaths."[43] But that's not what they meant to Post and his contemporaries.

Uses and Absences of Aerial Glacier Photographs

During his lifetime Post's photographs performed a variety of tasks. Often they were used to plan more-detailed studies, something that the great organizer, Bill Field, pointed out as their chief value.[44] They could be simply illustrative, such as those gathered into *Glacier Ice*, a large coffee table book that Post cowrote with snow scientist and fellow JIRP alumnus Edward LaChapelle in 1971. That book introduced scientific and popular audiences to the basics of glacier science. It was dedicated to the memories of Richard Hubley, whom they deemed "originator of aerial glacier studies," and William Fairchild, who died in 1969 during a takeoff from the Port Angeles runway.[45] *Glacier Ice* used Post's and others' photographs to illustrate glacier features and behaviors. There are pictures of eskers and ogives, crevasses and berschrunds, snowfields and termini, and much more. In this context, Post's photographs were beautiful illustrations that worked to supplement the written text.

But Post's photographs were more than just starting points and illustrations. To a trained eye they were also tools of prediction. He could discern a lot about glacier dynamics from a photograph. In 1974 Post used his photographic record of the Columbia Glacier to successfully predict that it was soon to drastically retreat (fig. 21). From the early 1980s to 2005 the Columbia retreated 15 kilometers.[46] Post forecasted this retreat six years before it began. He could use photographs to make predictions about glacier dynamics, where the Vauxes and even Field could not, because at this point glaciologists had better "on the ground" knowledge of the physics involved. It is likely that his singular ability to "read" glaciers also played a role. In one prescient yet unheeded insight, he pointed out the risk that icebergs calving from the retreating Columbia would pose to tankers carrying oil from the Trans-Alaska Pipeline. Fifteen years later the *Exxon Valdez* ran aground while attempting to avoid one such berg. More than 240,500 barrels of oil were dumped into Prince William Sound in America's largest marine oil spill until the Deepwater Horizon calamity in 2010.[47]

Finally, during his career Post's photographs were sometimes used to make estimates in lieu of quantitative studies. His aerials of firn lines—the border between bare ice and snow at the end of the melt season—were used alongside topographic maps to estimate a glacier's accumulation area ratio, which can be used as a proxy for mass balance.[48] Glaciologists of the 1960s made further at-

FIGURE 21 Oblique aerial perspective of Columbia Glacier, 1969. This image is very similar in composition to the one later used by Al Gore's team in *An Inconvenient Truth*. Photo by Austin Post. Glacier Photograph Collection, National Snow and Ice Data Center, Boulder, Colorado.

tempts to extract quantitative mass balance data from aerial photography, but as a discussion of the matter concluded in 1964, "in theory the method is very simple and appealing, but its application is fraught with many difficulties."[49] Aerial photographs revealed glaciers as surfaces; mass balance calculations required treating them as three-dimensional objects. Glaciologists created work-arounds that combined on-the-ground sampling and aerial perspectives, but these were imperfect approximations. Even today, remote sensing cannot fully eliminate the need for boots-on-the-ground measurements.

Post's repeat aerial photographs served a variety of purposes during his career, then. But they did not operate as harbingers of worrisome recession. Nowhere in *Glacier Ice* did the authors express concern over receding glaciers. While they recognized that the bulk of the world's known ice streams were receding, in the section on glacier fluctuations they devoted more words and pictures to the advancing ice. The exceptions, it seems, were more interesting than the rule. In the book's concluding paragraphs, LaChapelle and Post stated that knowledge of how glaciers respond to climate is critical for understanding "the earth-wide environment and man's [*sic*] place in it," but continued, "for we are [. . .] still in the midst of an ice age." As far as they were concerned, it remained to be seen whether the cycles of Pleistocene glaciations had come to an end.[50] This makes sense, since in 1971 few glaciologists believed glaciers were retreating indefinitely; some even believed another ice age was on the way.[51] Yet, even the reprint of the book in 2000 failed to mention the potential for its images of receding ice to be seen as evidence of global warming. While this may have been a matter of restrained science writing, it nevertheless suggests that these images were not yet seen as icons of ruptured geological time.

Post's work was the pinnacle of midcentury glacier photography, but he did not regard it as an exercise in witnessing global glacier retreat; nor did most of his colleagues. Initially, repeat glacier photographs were not seen as the canary in the coal mine, warning glaciologists that the planet was unnaturally warming. So what was the warning? Understanding how they entered scientific discussions about global warming requires first knowing a bit about the history of global warming research.

A Selective History of Global Warming Research

Post's career in glacier photography and the study of human-induced planetary warming developed in parallel during the latter-half of the twentieth century. The IGY kickstarted Post's aerial censuses; it also marked the start of an important dataset gathered at a laboratory in Mauna Loa, Hawai'i. There, climatologist Charles Keeling began regular measurements of the concentration of atmospheric CO_2 and found that it was on the rise. Plotted on a graph, the rise in concentrations of CO_2 traced a curve that swept swiftly upward in a steady arc that traversed a horizontal

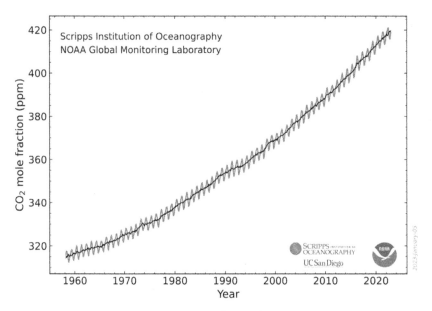

FIGURE 22 Concentrations of atmospheric CO_2 measured
at Mauna Loa, aka the Keeling Curve. Global Monitoring Lab,
National Oceanic and Atmospheric Administration

axis marked by years (fig. 22). Today this arc is known as the Keeling curve, itself
an icon of global warming, generating its own "culturally coded messages," as
historian Joshua Howe phrases it, of alarm and unidirectionality.[52] Greenhouse
gases like CO_2 and their role in the Earth's energy balance had long been known to
scientists, the grounds of which having been laid by Eunice Newton Foote, John
Tyndall, Svante Arrhenius, and engineer Guy Stewart Callendar in the nineteenth
and early twentieth centuries. Since then we've known that high concentrations
of greenhouse gases correspond to higher average global temperatures. But the
obstinate ascendency of CO_2 in Keeling's curve made the gas a new concern for
postwar scientists. It incited them to do more research.

Further impetus came from the work of climate modelers in the late 1960s and
1970s, who sought to understand and predict weather and climate through sim-
plified models of complex systems. Since global climate is such a "vast machine,"
involving many of the earth's systems, models are one of the only ways to study
how the various parts work together.[53] In the mid-1950s general circulation models

(GCMs) became one of climatologists' most important tools. GCMs aim to mimic the behaviors of global climate by accounting for planetary-scale phenomena like atmospheric circulation and oceanic currents. The initial GCM was built by Princeton physicist Norman Phillips using a liquid-filled dishpan and a primitive computer. Although the model "exploded" after twenty days, it managed to feasibly mimic a two-layer atmosphere. Phillips's attempt thus demonstrated the possibility of modeling the global atmosphere in three dimensions and spurred efforts to construct better models that could predict what increasing concentrations of CO_2 might do.[54] In 1967 Syukuro Manabe and Richard Wetherald of Princeton's Geophysical Fluid Dynamics Laboratory created a mathematical model of global atmosphere that showed a doubling of CO_2 over preindustrial levels would result in a 2.3°C rise in global surface temperature. In the years that followed, climate modelers strove to improve this estimate. They incorporated atmospheric aerosols and dust, variable humidity, seasonality, topography, and behavioral differences of distinct atmospheric layers into models run on increasingly powerful computers. The three-dimensional GCMs of the 1970s were a far cry from Phillips's dishpan. They convinced many scientists that increased CO_2 would lead to human-caused global warming.[55]

These modeling efforts and the persistent complexities and uncertainties involved in predicting the future of global climate caused the US government to take note. In 1977 a National Research Council (NRC) panel chaired by oceanographer Roger Revelle called for more research into the problem of CO_2. Two years later a second NRC report, the "Charney Report," was commissioned by President Jimmy Carter to look more closely at the question of human-induced warming and to evaluate climate models. The report concluded that the evidence that humans were contributing to changes in the chemistry of the atmosphere was "incontrovertible."[56] These reports prompted an increase in research on climate and greenhouse gases throughout the 1980s. In 1988 the Intergovernmental Panel on Climate Change (IPCC) was established to collect and assess scientific knowledge on climate change. The next year, in an all-too-brief flash of hope for American climate politics, President George H. W. Bush fulfilled a campaign promise by establishing a US Global Climate Change Research Initiative.

American climate research in the 1970s and 1980s grew out of efforts to better monitor and predict weather, to answer basic science questions about how the atmosphere and climate work, and to address concerns about global resources that arose against a backdrop of droughts, famines, and anxieties over global population. These are standard elements in histories of global warming that focus on American developments. But there is more to the story, even without widening the scope to include people and events beyond the United States. Historians of science have shown how cold war geopolitics shaped the science of global warming by encouraging scientists to rethink the globe as an object that could be known and monitored in its entirety. Alaska and the Arctic were not the only places military hawks wanted to transform into domains of prediction and control. The whole Earth lay within the bounds of their ambitions. Bolstered by the successes of the IGY and its successor, the International Hydrological Decade (1965–1975), their aspirations were furthered by the creation of international infrastructures, such as the United Nations Environment Programme's Global Environmental Monitoring Service (GEMS), and materialized in surveillance technologies like satellites—what historian of science Paul Edwards has termed "technological infrastructures."[57] The ambitions captured by these institutional and technological infrastructures were furthered by even greater leaps in computing power and modeling abilities. Postwar earth scientists redefined the "Earth as a single entity that could be investigated, measured, and represented as a single whole."[58] While they did not actually manage to control and predict all geophysical aspects of the planet, they made considerable headway in monitoring its resources and understanding how it works. What they learned in the attempt gave credence to the idea that human activity, such as widespread burning of fossil fuels, can alter planetary systems and pose environmental threats at the global level.[59] If we can know and monitor the whole Earth, can we not also transform it?

Whither Glaciology?

This brief and truncated history of global warming research stars computer programmers, oceanographers, and atmospheric scientists; glaciologists are thin on the ground.[60] Yet, this was the same time that the northwestern coterie including Meier, Hubley, and Post were reconceptualizing glaciers as hydrological resources and the targets of regional monitoring, that is, as environmental objects. Meier, especially, who was working for an agency committed to monitoring water resources for industrial and domestic use, was well-poised to give glaciology a place in emerging global monitoring research. Moreover, these northwestern scientists were not alone. Glaciology more generally was shifting toward the monitoring techniques that characterized the earth sciences' embrace of the global environment. The 1960s and 1970s saw the first efforts since the *Commission Internationale des Glaciers* (1894–1925) to establish worldwide monitoring of glacier fluctuations and mass balances. In 1973, as part of the International Hydrological Decade, the World Glacier Inventory was set up, overseen by representatives from the International Commission on Snow and Ice (established in 1948), the Swiss Federal Technical Institute, and the UN Environment Programme.[61] That same year the Swiss Institute established the Permanent Service for the Monitoring of Glaciers. Like the world's fresh waters, forests, and mineral deposits, glaciers were being viewed as environmental resources worth monitoring. This was inspired by a concern about freshwater resources and the desire to understand how twitches in climate might affect them. The transformations in the research agenda of northwestern glaciologists, then, was but one element in a turn toward resource monitoring and environmental thinking in glaciology more broadly, itself a fractal of trends in the contemporaneous earth sciences.[62]

But changes in disciplinary orientation and research agenda do not happen overnight. Despite these efforts, in the early 1980s, when other scientists and government officials were beginning to take seriously the possibility that humans were altering the climate, glacier monitoring efforts were still partial and incomplete. Comprehensive inventories existed for Alpine glaciers, but many of the world's then three hundred thousand small glaciers and ice caps were unaccounted for. North America posed special challenges because of the number, remoteness, and broad distribution of small glaciers and ice caps. Glaciologists hoped that remote sensing would help them fill in the blank spots on the map, but radar interoferometry and the gravity recovery and climate experiment (GRACE), two standard methods today, were still distant on the horizon. Even today, mountain glaciers pose unique obstacles to remote sensing. Areas of high relief, like mountains, can frustrate routine methods such as altimetry (measuring altitude) and gravimetry (measuring slight variations in gravity that correspond to different forms and features of the Earth). Optical satellites are often foiled by clouds and heavy snowfall, and radar is hampered by shadow, the effects of steep topography, and a rapidly changing environment.[63] Remote sensing, while promising, especially on relatively flat terrain like the ice sheets in Antarctica and Greenland, was unable to solve all the challenges of global glacier monitoring.

But what about repeat photography? Were Post's aerial censuses not evidence that global warming was happening? Many of the ice streams he photographed were indeed receding. Yet Post's glacier photographs did not simply proclaim "global warming is happening!" Appreciating why this was the case underscores the historical specificity of evidence.

What counts as convincing evidence in one context might not be convincing or even relevant in another. This is not a new idea in the historical and philosophical study of science. As early as 1979, philosopher of science Helen Longino argued that the relation between hypothesis and evidence—whether evidence counts toward or against a particular hypothesis—depends on background beliefs and assumptions. For our purposes, Longino's relevant insight is that some state of affairs can be taken as evidence for different, even conflicting, hypotheses. She illustrated this with a famous and instructive example from the Enlightenment: the discovery of oxygen. In the late eighteenth century chemists were trying to figure out how combustion and respiration worked. The British chemist Joseph Priestley made a series of experiments in which he burned various objects in a sealed bell jar and then inserted plants and living creatures to see how they fared. When a mouse and burning candle were placed in the jar, the candle was snuffed out and the mouse died. When a sprig of mint was added, a more fortunate mouse did not die and the candle continued to burn. He interpreted his results as evidence for the presence of what he called "dephlogisticated air," that is, air that had been deprived of its phlogiston by the action of combustion. Phlogiston was a substance that chemists posited to explain why things burned. During combustion, objects said to be rich in phlogiston released it and in doing so were reduced to ash. Objects that did not burn well were said to be phlogiston poor. Antoine Lavoisier, Priestley's Francophone colleague, rejected the phlogiston theory because it was unable to explain certain phenomena. When he performed the same experiments, he saw the same results as evidence for the presence of a hitherto unknown gas, which he called *oxygène* gas. This gas, he maintained, was consumed from the air during combustion. Longino concludes that what each man believed the results were evidence for depended on what else each believed about chemistry and his background assumptions about how the world worked. Evidence is context sensitive.[64]

So, too, with glaciologists in the 1970s and 1980s. While the possibility of global warming was being worked out, it was not glacier photographs that convinced scientists: it was evidence from climate modeling, paleoclimatology, geochemistry, oceanography, and atmospheric science. This is because they brought to their assessment of glacier photographs their historically conditioned beliefs about what constituted "good" evidence. Among glaciologists, the old criticisms of photographs as partial and inexact still held, as did the norms of precision and accuracy championed by geophysical glaciology that required knowing how much volume was being lost, not just whether the terminus had retracted.[65] While its influence on research agendas had been tempered by Hubley's critique, the geophysical turn had downstream effects for how glaciologists thought about evidence. Moreover, photographers like Post and Field, who had spent decades observing glaciers ebb and flow, knew too well glaciers' long, fluctuating geological histories. They were disinclined to draw what seemed hasty conclusions about comparatively recent trends of recession. Their background beliefs about geological histories of glaciers meant that they were unlikely to see repeat photographs of recession as evidence for fossil-fueled global warming. The photographic evidence captured trends that seemed too brief to support conclusions about long-term glacier projections. Knowing what we now know about the history of scientific glacier photography allows us to see how the twists and turns of this history shaped they ways glaciologists regarded the evidential role of photographs.

However, even though they may not immediately have seen global warming at work in photographs of receding glaciers, this does not mean that glaciologists were late to the consensus on global warming, as some historians and glaciologists have suggested. Certainly, some glaciologists, including a few working at Sweden's Tarfala Laboratory, resisted the idea that humans are responsible for rising global temperatures, well after most relevant scientists accepted the idea.[66] But by the early 1980s many North American glaciologists believed that human-induced climatic change could be affecting mountain glaciers. Meier's colleague, Wendell Tangborn, recalled 1980 as a year of significance: the year "the mass balances went negative."[67] Yet mass balance data was still spotty. So what brought glaciologists to scientific discussions of global warming? Glacier contributions to sea level rise.

Small Glacier Contributions

The first calculations of land ice contributions to sea level rise were made in 1940 by the Icelandic glaciologist Sigurdur "Skallagrím" Thorarinsson (nicknamed for the famous Viking farmer and skald of the sagas, Egill Skallagrímsson). Thorarinsson, Ahlmann's colleague and protégé, subscribed to his mentor's theory of polar amelioration, having witnessed extensive ice loss on Vatnajökull, which he, like Ahlmann, understood as part of nature's history, not a worrisome human intervention in the natural order. Yet, it wasn't Vatnajökull's decline that he sought to understand with his calculations of sea level rise. Scandinavian lands are quite literally on the rise. They are slowly bouncing back after having been depressed by the weight of tremendous ancient ice sheets. Earth scientists refer to this phenomenon of land bouncing back as isostatic uplift. In 1940 scientists noted unaccounted-for values in the isostatic uplift of Scandinavian coasts.[68] Thorarinsson thought that those values could be explained by a rise in sea level due to melting ice. The possibility that this melt could be human-induced was not on his mind when he calculated how much water glaciers were contributing to the sea in 1940. He sought to explain something else entirely.

The situation was different thirty-eight years later when John Mercer, a glaciologist at Ohio State University, predicted that rising levels of atmospheric carbon dioxide could lead to a rapid loss of ice in West Antarctica and a 5-meter rise in global sea level within a matter of centuries. This would entail doomsday scenarios of Noachian proportions; the streets of coastal metropolises would be inundated. Unsurprisingly, Mercer's paper got noticed and was republished in *Nature* as "West Antarctic Ice Sheet and CO_2 Greenhouse Effect: A Threat of Disaster."[69] But Mercer was a bit of an odd duck in scientific circles—rumors abounded that he did fieldwork in his birthday suit.[70] This might have contributed to why his paper provoked mixed reactions. Critics accused him of making dramatic speculations that were poorly grounded on weak evidence. Ice sheets like the West Antarctic Ice Sheet (WAIS) are dizzyingly complicated. Unlike glaciers, which flow in a single direction, ice sheets flow unhindered in multiple directions, like infinite spokes on a bicycle wheel. There were elements that Mercer could not account for, such as rapid deglaciation—when ice runs swiftly to its disintegration at the terminus. Such events were poorly understood, yet they are crucial for a solid grasp of ice sheet dynamics. Moreover, Mercer's predictions ran counter to what many scientists believed. Observational work beginning with the IGY suggested that WAIS was *gaining*, not losing mass. This doesn't mean that Mercer was wrong, or that he was doing bad science. Present-day glaciologists now share his concerns about the instability of WAIS, which in the early 2020s was pumping ice into the Amundsen Sea at a rate 77 percent faster than in 1973.[71] Still, in 1978 Mercer made bold conjectures based on thin evidence, and many of his colleagues thought him irresponsible.[72]

Though many of his scientific peers were skeptical of his calculations, Mercer's paper got them wondering about ice and sea levels. As Jessica O'Reilly has shown, it stimulated efforts to model the actions of WAIS, and prompted symposia and conferences, including one in 1979 at Canberra, Australia, and another at Orono, Maine, the following year.[73] The papers presented at these meetings emphasized the uncertainties surrounding global warming's potential effects on the cryosphere. Nearly every author called for more research.[74] The take-home objective thus became a review of knowledge about the processes governing interactions among sea level, ice, and climatic change. They did what scientists typically do in the face of uncertainty and patchy data: they called for more research.[75]

Attendees knew these gatherings were more than just intellectual discussions of a nifty scientific problem. They needed to produce a road map for future assessments.[76] But typically an assessment requires a target.[77] And there was a question of what target date they ought to set for future projections. Should it be fifty years hence? Two hundred? Behind closed doors, senior attendees at the Orono meeting chose the year 2100. Such a date was distant enough that policy makers wouldn't feel guilty if they didn't act immediately, but soon enough to make people worry and hopefully take action.[78] Policy implications were never far from scientific discussions of global warming, and scientists knew that navigating the political tightrope would be tricky.

WAIS dominated these meetings. On one level, this makes sense. West Antarctica contains over 3.2 million cubic kilometers of ice—enough to fill nearly 265 Lake Superiors. Most of it is grounded underwater, in some places 3 kilometers below the surface. Like tidewater glaciers that surge forward when their submarine anchors release, if the ice sheet's toehold on the ocean floor were to slip because of warming waters, rapid deglaciation would likely occur, sending titanic flat-topped bergs sailing into the Southern Ocean.[79] Still, the focus on West Antarctica is somewhat puzzling because glaciologists knew that small glaciers and ice caps would respond to warming trends more quickly than the ice sheets, and observation and theory for polar ice were weaker. Perhaps they thought it best to get to work on a big job that they knew would take awhile. Perhaps the romanticism of Antarctica, which was so potent during the IGY, still lingered. Or perhaps their imaginations were captured by the morbid scenarios that would ensue should WAIS collapse.[80] Mercer was not wrong that the results of any scenario would be disastrous. The streets of New York, Los Angeles, Lagos, and Sydney would be invaded by warmer and higher seas and Pacific atolls would disappear like so many Atlantises. Against such gloomy projections, the potential contributions of nonpolar glaciers, accounting for only 10 percent of land ice, might seem pithy.

Yet not everyone was fixated on WAIS. Mark Meier was at the Orono meeting and he was pondering smaller ice masses. When asked if he thought it possible that contributions from small glaciers and ice caps could explain unaccounted for rises in sea level during the twentieth century, Meier initially replied, "No, of course not, they're too small."[81] But the question planted a seed. After performing a few calculations, Meier changed his mind. Extrapolating from the mass balance records of a handful of glaciers (including the South Cascade), he landed on a figure that was remarkably close to that which needed to be explained.[82] Three years later he brought a more rigorously worked out figure to a meeting of the American Geophysical Union.[83] And the following year, in 1984, he steered a workshop in Seattle devoted to assessing the correlation of land ice changes to sea level rise. Drawing specialists in mountain glaciers, polar ice, permafrost, and snow, the meeting demonstrated how glaciology had grown since Odell defined it as the study of "living glaciers." The ice specialists were joined by climate modelers and oceanographers. Keeping the policy implications of their work at the fore, organizers stressed the need for an authoritative summary that could present a scientific consensus on the matter. Consensus had become an important goal for experts in the twentieth century, as they began to be more frequently courted by government to produce knowledge that could inform policy. It was a way for groups of scientists to "speak with a collective voice" to power.[84] The summary of the Seattle workshop highlighted the role of small glaciers in sea level rise: much remained unclear regarding the roles of Greenland and Antarctica, but Meier's work had shown that mountain glaciers and small ice caps were contributing to rising ocean waters. This rise would continue, the authors predicted, elevating sea level by a few tenths of a meter by the year 2100 and, they expected, the effects of a melting Greenland would also then be discernible.[85]

A few months after the Seattle workshop, Meier brought this work to the wide catchment of *Science* readers. His article "Contributions of Small Glaciers to Global Sea Level" put the issue of glaciers and global warming on the radar screens of many of his nonpolar colleagues.[86] His methods involved intuitive projections based on data from a handful of glaciers. Nevertheless, many found his attempt more credible than earlier predictions, including Mercer's, that wrestled with the conundrums of polar ice. The paper was auspiciously timed. In his referee report for *Science*, Bill Field wrote that "the subject is timely, as we may well be intering [*sic*] a period of increasing concern about the effects of CO_2 in the atmosphere." Mountain glaciers were a good entry point because, due to their relatively small masses, they would be among the earliest registers of slight climatic changes.[87]

Meier's work went on to form the basis of the section on small glacier contributions to sea level rise in the IPCC's first assessment report.[88] Formal and institutionalized, these reports are the gold standard for scientific consensus on global warming. While predictions of sea level rise have vacillated, due to the uncertainties involved in assessing WAIS and the need for IPCC scientists to present a univocal consensus, the contribution of small glaciers was far less controversial and unclear. By 2003 Meier and his coauthors were emphasizing the potential disappearance of small mountains glaciers, the contributions of mid-latitude glaciers to sea level rise, and the warming of arctic glaciers.[89]

The desire to understand small glacier contributions to sea level rise grew into a research program that considered how mountain glaciers were implicated in global warming. It provoked nonpolar glaciologists to join efforts to assess the severity and potential consequences of global warming. The foundation for this work was laid in the post-IGY conceptualization of glaciers as hydrological resources to be monitored on regional and, aspirationally, global scales. It lay, that is, in the orientation of environmental glaciology—a way of doing glacier research that treats glaciers as integrated elements in global systems, needing to be monitored like other natural resources. These efforts arose in the 1960s and 1970s when a new way of thinking about the globe as an interconnected, multiscalar whole began to emerge. Glaciologists communicated ideas in words and pictured them in equations, charts, and graphs. Repeat photographs did not feature prominently in the papers devoted to assessing sea level rise, even as the work became more and more focused on global warming. Here and there an Alpine glacier might grace the cover of a report, as the Rhône Glacier did for the Seattle workshop publication. But such visuals were predominantly aesthetic and illustrative, not evidential.

The paucity of repeat glacier photographs should not be surprising. Glacier contributions to sea level were *calculations*, communicated among scientists interested in precise changes in glacier volumes. The value of precision advocated by geophysical glaciology still held sway in professional circles. And these aims and values conditioned what scientists were inclined to accept as evidence. Austin Post's photographs were put in the service of understanding and predicting flow dynamics, in planning other glaciological research, and in getting a general sense of regional trends. But on their own they did not meet the standards of evidence articulated by Sharp's generation and still upheld among glaciologists. During the 1970s and 1980s repeat glacier photographs could at best be viewed as contributing evidence or a starting point for further studies. They were not evidence that global warming had likely begun. What counts as evidence depends on the historical conditions under which it is interpreted. Something needed to change for repeat glacier photographs to be viewed as evidence of global warming.

5

WITNESSING / *The Iconography of Ice*

To ask whether a photograph is analogical or coded is not a good means of
analysis. The important thing is that the photograph possesses an evidential
force, and that its testimony bears not on the object but on time.

ROLAND BARTHES / *Camera Lucida*

Bruce Molnia has been taking pictures of ice since he was a college student. During
a phone call in 2018 he told me that he was first inspired by the seductive lines and
tones of Antarctic ice while visiting the southernmost continent on a break from
school in 1965. His eye for the aesthetics of ice followed him through a fifty-year
career in government science, and he photographed glaciers simply "because they
were interesting geological features." Molnia is an expert in marine geomorphology
and glacial sedimentation. Most of his work has focused on Alaska. As he wound
his way through the offices of the USGS, the National Research Council (NRC),
and the US Congress, he became finely attuned to the value of making science
serve publics.[1] In 1999, repeat glacier photography became part of his job. That
year, with climate change denial on the rise, he was charged by his superiors with
finding evidence of global warming that was convincing and easy to understand.[2]
What could be more accessible and easy to produce than repeat photographs of
places he knew from his time working in Alaska? "The simplicity of the photos is
so striking," he wrote, "my basic premise is, if a picture's worth a thousand words,

what's a pair of photos showing dramatic change worth?"[3] For the next five years he re-created over two hundred historic glacier photographs, reoccupying the photo stations set by Bill Field, Harry Fielding Reid, Charles Wright, and Grove Karl Gilbert, and re-creating photographs he had taken as a young geologist in the 1970s. His work is careful, precisely re-creating the scenes of his predecessors. His photographs tend toward the monochromatic; some have a cerulean beauty of indigo glaciers and azure waters against sapphire skies; others wallow in the grays of wasted ashen glaciers and silver waves backed by pallid clouds. Often his repeat images elicit contrasting associations, between vigor and disintegration (see figures 23 and 24). This work was highly successful and appreciated by his peers. In 2011 he received the USGS's Eugene A. Shoemaker Award for Lifetime Achievement in Communications, for "exceptional contributions [. . .] communicating USGS science to non-scientific audiences ranging from school children to supreme court justices."[4]

Molnia was part of a larger trend. In the early 2000s glacier scientists resumed taking repeat photographs of ice. They were joined in this venture by landscape photographers, journalists, and recreationalists, all of whom also turned to repeat photography as evidence showing that the cryosphere was disappearing at alarming rates.[5] For the glaciologists it was a return to something they had largely set aside. One hundred-odd years before, glacier naturalists used repeat photography to study glacier fluctuations as they scrambled among North America's western ranges. After the Second World War, glaciologists de-prioritized photographic documentation of termini fluctuations, turning instead to instruments like the seismograph and ablatograph to analyze the physical structures and dynamic processes of ice and snow. Certainly those with an eye for alpine beauty continued to photograph ice and snow. Bill Field never stopped photographing glaciers, but this caused him to sometimes be regarded by colleagues as "not *really* a glaciologist," a claim this intensely modest man would likely have agreed to.[6] Even in the 1960s and 1970s, aerial repeat photographer Austin Post worked on the margins of his science, struggling to secure funding, being well-respected by some but arguably better appreciated by those who came later. Many repeat glacier photographs of this period shared the fate of ones taken by Maynard Miller on the Juneau Icefield: to lay unsorted in boxes, jumbled among snapshots of mountaineering escapades and career high points.[7] Tracking down collections that are not yet archived sends

FIGURES 23 AND 24 Repeat photographs of the Muir Glacier
by Bill Field in 1941 and Bruce Molnia in 2004. Glacier Photograph Collection,
National Snow and Ice Data Center, Boulder, Colorado.

historians like myself on meandering chases that often end in cluttered garages, basements, or office closets. These photographs were understood to be of less scientific value than the data acquired through more quantitative means. Beginning with the geophysical turn in the middle of the century, repeat glacier photographs like Post's were typically starting points for other investigations that met rigorous standards of precision and quantification. Why, then, did scientists like Molnia take up repeat photography again, seemingly eschewing these entrenched ideas about good evidence?

The answer requires thinking about science as a historian does. This is something we have been doing all along but about which we should now be more explicit. The historian's science is not an abstract body of accumulating knowledge; rather, it is science understood as a nexus of evolving practices, materials, institutions, people, and sometimes other critters, like lab rats or nematodes, embedded in various societies and cultures at particular moments in history.[8] In the three regimes of glacier study considered here—glacier naturalism, geophysical glaciology, and environmental glaciology—when knowledge-makers used repeat photographs as evidence, their concerns were epistemic. That is, they were related to producing and validating knowledge and were generally addressed to scientific colleagues. The return to repeat photography in the early 2000s was otherwise motivated with a different set of nonepistemic goals: objectives not directly tied to the production of knowledge. For university researchers, nonepistemic goals may include practical ends like securing grants or achieving tenure. They may be value-driven, as when scientists choose projects that allow them to do fieldwork in places they love or when they seek to produce knowledge that is relevant to policy. Nonepistemic goals are not extraneous, that is, not *merely* practical or political. They are an ineliminable aspect of how science gets done. They are an important part of the historian's science, as critical to the daily doing of science as are models and theorems. And they impact the choices that scientists make, as was the case for glaciologists at the turn of the twenty-first century.

Glaciologists returned to repeat photography because they wanted evidence that could speak to audiences outside of their profession, what they understood as public audiences.[9] At the turn of the twenty-first century certain glaciologists became concerned that many nonscientists did not understand the facts of global warming. They turned to repeat photography as a form of public-facing evidence. Another way of putting this is to say that the nonepistemic context of science shaped their decisions about evidence and audience.

This does not mean glaciologists compromised their scientific integrity or did bad science. Such conclusions rest on the belief that nonepistemic elements are antithetical to science. This is not the historian's science. Rather, the return to repeat photography simply illustrates, again, that evidence is limited and context-dependent. No one type of evidence can do everything; scientists must choose the right kind for the job. And their choices are steered by epistemic and non-epistemic concerns that are unique to the historical conditions in which they work. Repeat glacier photography seemed to be a good type of public evidence of global warming. And in many ways it was. I, for one, can be counted among those who came to have greater concern about the effects of global warming by looking at repeat photographs of receding ice. But, as with earlier instances of repeat glacier photography, the legacy of these images turns out to be knottier than their simple format belies. Appreciating why folks like Bruce Molnia turned again to repeat photography after decades of relegating it to the margins of scientific practice requires appreciating the situation they found themselves in. This demands some background in the history of climate science denial. It is a convoluted tale involving malicious promulgators of doubt, unwitting journalists, and well-intentioned but tragic scientists.

Global Warming as a Problem of Perception

By the late 1990s most scientific experts agreed on the basics of global warming: greenhouse gases produced by burning fossil fuels were changing the chemistry of the atmosphere, causing the planet to warm and the climate to change rapidly and wobble unpredictably. There were those who wanted to know more about, say, the capacity of the oceans to absorb energy and convey it downward into deep circulation belts; there were (and remain) questions about future impacts; and there were scientist "mules," who dug in their heels and simply refused to budge from a position of skepticism.[10] But projections from increasingly powerful and sophisticated models were converging with observational evidence, pointing to the conclusion that humans were changing the energy balance of the planet and this was affecting climate. As the IPCC famously phrased it in its second assessment report, "the balance of evidence suggests a discernible human influence on global climate.[11] In plain terms: scientific experts believed humans were changing the climate. This was in 1995.

The bulk of these experts—persons accredited with specialized knowledge relevant to understanding the climate system—agreed that global warming was real and its beginnings could already be discerned. Many of their scientific peers respected their expertise in this matter. The trouble was that many people outside the scientific community, especially in the United States, didn't seem to accept this conclusion or to care that much when they did. In the first decade of the new century, American public opinion on global warming dipped and bobbed like a pro boxer in the ring. According to Gallup polls—which have been recording Americans' opinions on global warming since the late 1980s—concern about climate change peaked in 2000, when 72 percent of those surveyed said it was a matter about which they were concerned. In 2004 only 51 percent expressed concern. In 2006 a *New York Times Magazine* poll found that only about half of Americans even believed global temperatures were rising.[12] Then 2010 and 2011 saw new lows. Americans were getting less worried about global warming, getting less convinced that its effects were happening, and more likely to believe that scientists were themselves uncertain. A full 48 percent touted the belief that the seriousness of global warming was being exaggerated, a jump from the 31 percent who had thought so when Gallup first posed the question in 1997. For those who knew the planet was warming at unprecedented rates, which could only be explained by growing concentrations of human-generated greenhouse gases, this change in perception was a worrying situation.

It is worth noting here that the issue of global warming is not a black-and-white matter of believers and skeptics, the enlightened and the benighted, the faithful versus the pagans. There are many reasons why a person or a group of people might not endorse global warming as the cause of this or that observed phenomenon.[13] Postmortems of climate agreements like the Kyoto and Paris Accords reveal that the reasons people reject robust climate action go beyond a lack of understanding of the science. For instance, long histories of exploitation, punctuated by disagreements over land use, mean that Indigenous and chronically disadvantaged communities often have little reason to trust either scientists or environmentalists. As M Jackson and Karine Gagné have shown in their work on communities living near glaciers, historical precedents and cultural beliefs influence whether or not people understand glacier recession to be caused by global warming. They may regard it instead as a moral failing of the community to properly care for their glaciers, or they may see it as welcome change in a long history of glaciers wreaking havoc on their homes and farms. Such beliefs operate unevenly within communities. Not everyone in a single mountain hamlet sees the melting of the neighborhood glacier in the same way.[14] We disagree about global warming because it is more than just a physical phenomenon knowable through the natural sciences alone. As Cambridge geographer Mike Hulme explains, it is an idea that we, as diverse human populations, engage with through "our different attitudes to risk, technology and well-being; our different ethical, ideological and political beliefs; our different interpretations of the past and our competing visions of the future."[15] This, he argues, is why we need humanists and social scientists to be part of conversations about global warming. They can help us see how things came to be and maybe what can be done.

Years of scholarship have led to this nuanced understanding of global warming. Yet for scientists and environmentalists in the late 1990s, the phenomena captured in the polling numbers seemed like a straightforward problem of science communication, solvable by better efforts on the part of scientists to share their knowledge. And while the story is indeed more complicated, scientists weren't wrong to think that expert communication was part of the problem. Without scientific aids, global warming is largely invisible.[16] Certainly, people are already perceiving its touch: Inuit, Iñupiaq, and Inughuit Peoples residing in the circumpolar north, for example, are experiencing the effects of a rapidly warming Arctic. Transportation, hunting, and fishing have become more difficult as sea ice and permafrost are rendered less predictable and stable. Californians are experiencing the devastating culmination of decades of heat and drought in the wildfires that now hammer the state every summer. Yet, definitively identifying these experiences as fallout from *anthropogenic* climatic change requires the sophisticated instruments and models of scientists and an ability on scientists' part to communicate effectively. Indigenous peoples and nonscientists do not need scientists to speak on their behalf, but scientific experts are necessary for the identification and articulation of global warming as human-caused. It seemed to many in the first decade of the new millennium that scientists were failing to make a compelling case.

Making a compelling case was not as easy as it sounds. By the late 1990s it was becoming clear that the fight against global warming was not going to be simple.[17] Environmentalists and scientists were facing increasingly sophisticated political attacks on the science behind global warming. There were many reasons for this. Politically conservative politicians and scientists, motivated by ideological commitments to the free market, used tactics of doubt-mongering to cultivate public uncertainty about the facts of global warming, the potential threats involved, and whether anything could be done about it. Through the creation of bogus institutes, the circulation of white papers that only resembled peer-reviewed science, and the manipulation of evidence to suggest nonanthropogenic explanations, they fought credible science by aping many of its most recognizable features. These conservative scientists were well-practiced in promulgating doubt, for they had honed their tactics in earlier battles against research showing the harmful health effects of tobacco. "Doubt, is our product," a tobacco executive infamously declared in 1969, "since it is the best means of competing with the 'body of fact' that exists in the minds of the general public."[18]

These doubt-mongers exploited an unforeseen Achilles' heel in modern relations between expertise and policy making. Organizations such as the IPCC assumed the efficacy of what historian Joshua Howe has termed "the forcing function of knowledge"—the common-sense idea that if people have the right knowledge, they'll do the right thing. Thus, if scientists got the science right, the politics of emissions reduction and regulation would simply follow. It did not take long for political opponents to spot a weakness. The best way to influence the *politics* of global warming was to challenge the *science* behind it. Undermine the foundation and you undermine what follows. By relying first and foremost on science as a driver of political action, climate advocates exposed science as a target instead of inviting open conversations about the values involved. Howe diagnoses this as a tragedy of Ancient Greek proportions: the means by which scientists and activists sought to achieve their ends became the instruments of their undoing.[19]

Magnifying the tragedy, this undoing was inadvertently furthered by the media. A few science journalists were duped by the sophisticated, science-aping tactics used by cultivators of doubt. At a more general level they unknowingly supported the credibility of doubters by presenting global warming as a debate with two legitimate sides. The source of this misguided ecumenicalism was deeply rooted in the professional norms of their craft. Journalists, like jurors and scientists, (ideally) strive to be objective. Yet each profession has its own ideas about what objectivity means. The press (again, ideally) seeks objectivity by being a neutral, unallied fourth estate. We can see how this could make sense in the coverage of political news. In the American case, to remain neutral means giving all political parties a chance to air their views on an issue. The key to neutrality is balance. This principle is called the fairness doctrine, and it undergirds the profession.[20] Yet the fairness doctrine is ill-suited to the science of global warming (and science, generally), because the science is a matter of expertise, not a matter of political opinion. Certainly, there are political aspects to global warming, not the least of which is the daunting question of what to do about it. But the science is not a matter of campaigning; it is a matter of experts reaching agreement based on evidence and attested professional practice. In attempting to be objective as best they knew how, science journalists covered "both sides" of a scientific "debate" on global warming, even when one side represented the vast majority of relevant experts and the other only a small population of dissenters. This off-kilter coverage gave credence to the notion that global warming was uncertain and contested; only a mere—and controversial—hypothesis.

Pernicious doubt-mongering and the practices of science journalism led many nonscientists in the 1990s and 2000s to believe that there were two credible sides to the scientific story on global warming and substantial uncertainty among scientists. Yet scientists themselves tilled the ground upon which the seeds of doubt sprouted. Scientific communities, as *scientific* communities, are cautious about endorsing new claims. This is how they vet unsubstantiated claims and ensure that what is stamped as scientific knowledge meets the community's rigorous standards of proof. Scholars have identified this as a tendency toward explanatory *conservatism*—the propensity to shy away from what seems radical or exceptional.[21] The earth sciences, especially, come by this propensity honestly. Historians have traced it to the influential Victorian geologist Charles Lyell, a foundational figure of modern geology. He forcefully advocated for a science of the Earth based on uniformitarian principles—that is, principles that operate throughout the planet's history. Gradual, accretive processes, he believed, were more amenable to scientific understanding, in part because one of the other explanations for geological phenomena on offer at the time was the Biblical flood. Capricious acts of a remorseful god wishing to start over were not the most solid foundation for a predictive science. While the subsequent history of geology, with its ice ages and meteoric impacts, has tempered its Lyellian tendencies, the predilection for conservative explanations lingers among many earth scientists, making them wary when evaluating claims of novel, unprecedented change.[22] We've glimpsed this in Post's and Field's hesitancy to see the recession captured by their glacier photographs as something novel and human-caused.

By avoiding hasty conclusions and "erring on the side of least drama," scientists were simply trying to be good objective scientists. Unlike in journalism, where objectivity means balanced neutrality, in science it has come to mean the removal of the subjective, an absence of opinion and personality in the making and presenting of facts.[23] Drama, as feminist scholars have observed, seems antithetical to this.[24] Recall John Mercer's colleagues' wariness to endorse his paper on the disintegration of the West Antarctic Icesheet simply because he appeared to be indulging in dramatics. Along similar lines, in 2018 Bruce Molnia told me that when he saw colleagues make "radical statements," which were then proved wrong, he did not wish to share their fate, so he initially hesitated to make strong statements about glacier recession and global warming.[25] More than other social groups, scientists are jittery about "crying wolf" or being perceived as radical. Erring on the side of less drama may not have directly fueled uncertainty about global warming, but it did not actively counter it either. Inadvertently, opting for caution contributed to conditions amenable to promulgating doubt.[26]

The net effect of these deliberate and unintended contributions to doubt was that in the late 1990s and early 2000s, sundry nonexperts believed that the science around global warming was confused and uncertain. Scientists felt compelled to do *something* in response. They sought ways to share knowledge with people outside their professional silos. When it came to glaciers, one way forward lay in setting aside the precise, quantitative instruments and hard-earned markers of professional expertise developed in the mid-twentieth century and reviving an older form of evidence that seemed right for the job.

The Revival of Repeat Photography

Experts use arcane words that nonexperts don't understand. It's part of what makes them experts. Mastery of specialist vocabularies built especially for a group of trained inductees is a hallmark feature of modern expertise. These specialist languages, both discursive and visual, create and maintain borders between science and nonscience. They do what historians of science call "boundary work."[27] As such, they are productive and occult, generating reams of discourse among initiates that are inscrutable to nonexperts. Nonexperts may lack the technical skills required to "read" a representation, making it intellectually inaccessible. Alternatively, even when understood intellectually, representations may fail to compel a sense of meaning, rendering them emotionally inaccessible.[28] The ability to read charts and graphs, to decipher models and equations, or to translate the often dense prose of peer-reviewed articles are skills that require deliberate training. Without this training, such modes of communication may simply perplex or fail to inspire a sense of investment or empowerment. When it came to evidence of global warming, geologically trained photographer James Balog put the problem plainly in the film *Chasing Ice*: "The public doesn't want to hear about more statistical studies, more computer models, more projections; what they need is a believable, understandable piece of visual evidence, something that grabs them in the gut."[29] Evidence had to be intelligible and it had to generate a response. In the face of public confusion and uncertainty over global warming, scientists called for new forms of evidence.

They felt they needed new kinds of evidence because what they used among themselves couldn't do the job. Since the geophysical turn in the mid-twentieth century, of which glaciology was a part, earth scientists tended to work with quantitative representations of data. But this, as Balog suggested, was unhelpful for reaching nonscientists—not because such evidence was not good, but simply because it was limited, as all representations are. A single type of representation can't do it all and there are trade-offs to be made when choosing among different types. The simplest representation will not be the most complete, nor will it be the most useful for making predictions.[30] On the other hand, representations that strive for utter completeness, like Luis Borges's fanciful "imperial maps" and their exact point-to-point replication of a territory, cease to represent and instead recreate the thing itself—meanwhile becoming useless as maps. Less fantastically, representations that are too complete sacrifice predictive power and simplicity.[31] Scientists must use different kinds of representations to achieve different ends. The kinds of representations they use as evidence will change according to the audiences they seek to address. In the case of glaciologists at the turn of the twenty-first century, they were looking for evidence that would speak to audiences outside of the professional circles that adhered to standards of precise quantitative evidence. They needed evidence that would perform nonepistemic work for them when addressing nonscientist publics. They didn't need precision; they needed something accessible, and they needed something compelling. Enter repeat glacier photography.

In 1999, "when the first wave of climate deniers was surfacing," Bruce Molnia was challenged by David Hayes, counselor to the interior secretary, to find "unambiguous, anecdotal, easily understood documentation that climate change was real." He turned to repeat photography.[32] Mindful of his intended audience, Molnia released his photographs into the public domain so they might be freely shared to spread the message. Global warming was happening. It could be seen in the rapid disintegration of glaciers, subjects both exotic and beautiful. Molnia was a careful photographer. His re-creations closely mirrored the scope and composition of the originals, minimizing "noise" and making the changes to glacier fronts pop with a powerful lucidity. His intuitions about the persuasiveness of glacier photography proved well-founded. His photographs were widely taken up by news agencies, including a London-based company, Still Pictures Ltd., which brought them to high-profile meetings on global warming. They have been used by NASA to create an "Images of Change" app, they have adorned fridge magnets, they have been displayed at institutions like the Canada Science and Technology Museum, and they were the basis of a card game in which players have to match historic and recent photographs of glaciers—a sobering version of the game Memory for the twenty-first century.

Molnia was not alone in recognizing the potential of repeat glacier photographs as public evidence of global warming. In 2007, *National Geographic* devoted an entire issue to the cryosphere. The most popular article was the award-winning "The Big Thaw," a piece written by Tim Appenzellar—a heavyweight among science journalist who cover environmental topics—and illustrated by repeat photographs of receding glaciers taken by Balog. Balog is known for his starring role in the 2012 film *Chasing Ice*, a tale of redemption and hope in which he, a born-again climate champion, heroically devotes his life to spreading the gospel of receding ice (whatever the cost to his aging knees). "The Big Thaw" was Balog's slingshot to stardom. It opened with a two-page spread of a sapphire river meandering through an ice canyon on the Greenland Icesheet, followed by repeat photographs of Bolivia's Chacltaya Glacier (which once supported a high elevation ski resort but in 2005 was barely enough of a snow patch for tobogganing), Iceland's Sólheimjökull, and the Grinnell Glacier. The Grinnell, in Montana's Glacier National Park, was once one of the most recognizable glaciers in the Lower 48; today it is a smear of ice barely deserving the designation "glacier." "The Big Thaw" was one of *National Geographic*'s most successful articles of the mid-2000s. It launched Balog into an ongoing project to document receding glaciers around the world: the Extreme Ice Survey (EIS).

Balog is not a glaciologist, but EIS began with his collaboration with glaciologist Tad Pfeffer. One branch of its lineage, then, is Pfeffer's scientific work. A University of Washington graduate, Pfeffer inherited Austin Post's and Mark Meier's dynamical studies of the Columbia Glacier. Like Post, Pfeffer used photography in his studies of the glacier's dynamics. Before he met Balog, he had been using low-grade time-lapse photography to record movements on the Columbia's surface. Photography captured a piece in the puzzle of convoluted ice dynamics. It was a way of "looking up from" specialized measurements and broadening his purview to grasp what a myopic focus on the quantitative might miss.[33] The legacy of Richard Hubley's critique of geophyscial glaciology made itself felt in Pfeffer's approach.

Pfeffer came to the study of glaciers out of a sheer love of snow and ice, and he seems to have never lost that broader appreciation. Attuned to the beauty of the cryosphere, he is one of those experts who wants to share his work with nonscientists and is remarkably good at it. He has called for glaciology to descend from the removed and lofty heights of the ivory tower.[34] In 2007 he published a popular book on the history and current study of the Columbia Glacier. Similar in format to LaChapelle and Post's *Glacier Ice* (to whom Pfeffer acknowledges inspiration), it is a generously illustrated coffee table book that deftly explains complex glaciological phenomena in direct, engaging prose. With experience in photography and an impetus to share beyond the silo of his expertise, Pfeffer was well-suited to contribute to efforts to engage nonscientists.[35]

During the Third International Polar Year (IPY) (2007–9), Pfeffer attended a meeting sponsored by the National Science Foundation about sharing the IPY with nonscientific publics. A sense of gravitas infused the gathering, as participants sought ways to communicate the urgency of human-induced cryospheric recession to publics that seemed evermore doubtful. Pfeffer learned of James Balog's intention to use glacier photographs to reach skeptical publics. A fellow resident of Boulder, Colorado, he offered Balog his expertise. The two came together on Columbia Glacier in 2008. According to Pfeffer, Balog had been struggling with how best to use photography to document glacier recession. Initially Balog had thought to use a mosaic technique that he developed on a previous project photographing trees. He had used repeat photography in "The Big Thaw," but Pfeffer's technical know-how with time-lapse photography, gained from working on the Variegated and the Columbia Glaciers, put EIS on track to take automated time-lapse photographs of glacier fronts.[36]

Today, with sponsorship from Nikon, EIS operates time-lapse cameras in Alaska, Greenland, northern Canada, Iceland, the Alps, Antarctica, and the Rockies. These cameras can reveal months of recession in a matter of seconds. In the still-frame context the images are presented as repeat photographs. By automating the photographic work, images are captured all year round, in all conditions (even in weather that obscures the glacier). But it doesn't remove the need for the EIS team to travel to remote locations every year to retrieve memory cards and service their fleet of solar-powered Nikon D200 digital cameras. Stories of EIS fieldwork have been featured in films (*Extreme Ice* [2009] and *Chasing Ice*), books (*Extreme Ice Now* [2009] and *Ice: Portraits of Vanishing Glaciers* [2012]), websites, and magazines. These stories tend to follow a narrative pattern of heroic enlightenment in which photographers battle the elements to bring to others knowledge out of wild places.[37] In this context, glacier photographs demonstrate the urgent reality of global warming. They are part of M Jackson's "narratives of ruin."[38]

The work of Molnia, Pfeffer, and Balog was part of a turn toward repeat glacier photography in the twenty-first century that included scientists, photographers, and alpine recreationalists. Many others followed this pattern, combining science, art, and extreme recreation under the auspices of corporate-sponsored awareness raising. Some projects seem to use glacier photography as a veneer over what are at root mountaineering or skiing endeavors. Others have more robust scientific credentials. Glaciologists and climate change researchers such as Mauri Pelto, Heinz J. Zumbühl, Samuel U. Nussbaumer, Daniel Steiner, and Patrizia Imhof, for instance, have used photography to spread the message that global warming is melting the world's glaciers. So has the website Glaciers of the American West, which operates under the supervision of Portland State glaciologist Andrew Fountain, with support from the USGS, NASA, and the National Parks Service.[39]

These projects are premised on the idea that repeat photographs of retreating glaciers make global warming visible. Rising average temperatures aren't something you can see directly, but as Michael Zemp, lead scientist at the World Glacier Monitoring Service, told *National Geographic* in 2006, "a glacier melting is something everybody can see." Repeat glacier photographs seem to anchor the gargantuan, invisible issue of global warming in something concrete, visible, and aesthetically striking. You don't need special training to understand the visual logic of before-after, as you might for reading a mass balance chart. We're already versed in such imagery. This doesn't mean that repeat glacier photographs are transparent or free of ambiguity, but simply that their format is more accessible than the other kinds of evidence that glaciologists use. They help the science of glaciology, which—like all modern sciences—has become specialized, technical, and abstruse, take a step out of the ivory tower. "For all our emphasis on models and math," quipped Ted Scambos, one of Pfeffer's colleagues at the University of Colorado, "seeing is still believing."[40]

Access and Inspiration

After largely abandoning repeat photography in the mid-twentieth century, glaciologists returned to it fifty years later to provide to nonexperts what they believe was simple, compelling evidence of global warming. These are social and political concerns. It didn't make them bad scientists, or traitors to their discipline. Arguably, it makes them better, more well-rounded experts. It means simply that they chose to privilege certain representational traits over others. Understanding such choices requires taking into account the intentions of the representer: what the representation is intended *for*. Histories like this one help uncover what those motivations might be. Repeat photographs are not the best evidence, by the epistemic standards, handed down from geophysical glaciology. They are imprecise and even sometimes misleading. But given the social and political situation of the late 1990s and 2000s, they seemed to be the right evidence with which to communicate global warming to publics who seemed confused, uninformed, or misinformed. Accuracy and precision were, in that instance, traded for accessibility and inspiration.

Glacier photographs seemed to be a good choice as accessible evidence because they appeared simple compared to more technical forms of evidence. Photographs, more so than other visual representations, appear like clear mirrors onto the world itself. Contrasted with, for example, heat maps, with their splotches of yellow and red splattered across flattened geographies, photographs look more like the world we see. Of course, photographs are not *actually* transparent windows onto the world; a lot of work goes into their production, which affects how and what they portray. But thinking this way is typically the business of scholars and artists; most people operate as if photographs unproblematically capture reality.[41] Moreover, when reintroduced, repeat photography was familiar to viewers. Its before-after visual logic permeated the advertisements that still pepper the mediascapes of our lives. And it is indeed an effective way of conveying change over time. Repeat photographs highlight change because, against a fixed background, small alterations pop out. Media scholar Edward Tufte calls this the principle of "the smallest effective difference."[42] People are very good at detecting small differences against a backdrop of overall similitude. Thus, at least on the face of it, repeat photographs take less training to grasp than, say, seismographic readouts. Assumptions about photographic realism and familiarity with the form of repeat photography make it seem an accessible form of evidence.

As for inspiration, glaciers are an alluring subject matter. Since the Romantic movement tore through Western culture in the eighteenth and nineteenth centuries, mountain glaciers have been representatives of sublime wilderness celebrated for their "aesthetics of infinite." They dwarf the viewer and inspire a fearful, reverent awe.[43] Romantic visions of wilderness in North American landscape painting and wilderness photography have tended to emphasize grandeur and isolation, proffering visions of an unpeopled, ahistorical, and utterly magnificent nature. This vision of wilderness is deeply entrenched in environmentalist and preservationist portrayals of nature. Beyond aesthetic appeal, this is because wilderness encapsulates a potent politics that most of us accept without thinking: if wilderness is nature outside the human realm, then nature occupies a position of neutrality with respect to our political squabbling. This is a very old idea in Western culture, rooted in even older ideas about divine authority.[44] It underlies the desire to use photographs of glaciers to address the politics of global warming skepticism. Drawing a line between riotous human politics and apolitical nature, paleoclimatologist Lonnie Thompson maintains that "glaciers don't have a political agenda, they just kind of sum up what's going on out there and they respond to it."[45] This thinking makes glaciers desirable witnesses in media coverage that, as noted, had taken on the guise of a debate. Like the neutral journalist, glaciers' testimony was trustworthy because they were nonpartisan. Far from population centers and political corridors, glaciers are aloof and above human bickering: the ideal unbiased witness. As Thompson's remark suggests, glaciers are potent players in the politics of global warming precisely because they appear to be external to those politics.

More than that, like the ideal scientist, repeat glacier photographs lack the stamp of the personal. Another way of putting this is to say that glaciers are understood to be natural laboratories, and the camera a passive instrument of reproduction that, together, are capable of automatically registering global warming in the fluctuations of glaciers' bodies. Repeat glacier photographs appear to be products of self-reporting instruments and so eliminate the need for interpretation, which can be subject to personal bias. Glaciologists like Molnia and Thompson did not say this to themselves when they turned to repeat photography. They didn't have to. It is woven into the way that many of us, including scientists, think about wilderness landscapes: as outdoor laboratories from which Nature (capitalized for authority) speaks.[46] Unspoken cultural assumptions about photography and the cultural and epistemic authority of wilderness underwrites the appeal of repeat glacier photographs as public-facing evidence. They are both accessible and compelling.

Nonlinear History

This account of the fortunes of repeat glacier photography in the twentieth century reveals that evidence has a history.[47] More accurately, it has many histories. The one related here is not *simply* a tale of increasing technicality and abstruseness, where the end is fated by the intrinsic logic of science. That is, there is nothing preordained about the development of scientific evidence, though we might assume it always gets more technical. Practicing scientists, "situated in time, space, culture, and society, and struggling for credibility and authority," pursue a variety of goals in their job to produce and share reliable knowledge.[48] These change over time and are responsive to political and social contexts. They rely on often-unspoken beliefs about what counts as proof and what matters. When viewed in context, we see that there is no such thing as *best* evidence independent of the relationship between presenter and intended audience. Repeat photographs do not count as particularly compelling evidence for many glaciologists who, as heirs to midcentury professional standards, seek instead more precise and predictive representations. But for nonexperts, for whom mass balance charts are all but unintelligible, repeat photographs of receding ice operate as compelling and seemingly unambiguous evidence that humans are changing the global climate.[49] This is what occasioned the return to repeat glacier photography that populates today's iconography of ice, offering a vision of global warming as a threat to distant, innocent, pristine lands.

CONCLUSION / *People and Glaciers*

I began researching and writing about glaciers because I wanted to get out of graduate school. In 2013 I was, as many others have been, frazzled and miserable during the first year of my doctorate program. I was far from home in a big, frenetic American city at an institution that made me feel constricted and uncertain. And it was hot! A summer in Boston would make hell feel cool.[1] I was a hair's breadth away from quitting school when chance intervened in the form of an off-hand comment from a professor: "If I could give graduate students one piece of advice, it would be: study what you love."[2] The rest of the conversation never made it to my ears. I was gone, immediately roving in my mind's eye over peaks and icefields. I could study glaciers, because what did I love more? I thought of *Chasing Ice*, which had just come to theaters and inspired in me a new concern for the fate of the cryosphere. The time lapse and repeat photographs told a compelling story. I thought of the Bow Glacier, and of the black-and-white photograph hanging in the lobby of Bow Lake Lodge. What was the story behind that image?

This book is proof that glaciers can inspire strong feelings of love and attachment, even from far away. Early glaciologists seeking accessible and compelling evidence of global warming had plenty of reasons to turn to repeat glacier photographs. And repeat glacier photographs are generally a pretty good record of landscape change over time. Yet the choices scientists make about what to use as evidence impacts how the phenomenon in question is understood. Visualizations affect how we see the world, making certain aspects visible and rendering others unseen. The historian's science lives on a two-way street: not only do the social, cultural, and political influence the mundane doing of science, the practice of science has social, cultural, and political effects. This was certainly the case for repeat glacier photography. Largely set aside after the Second World War, it returned at the turn of the twenty-first century. Glaciologists circled back to an older form of evidence that seemed to offer new value in light of climate change denial. Those who have turned to repeat glacier photographs as evidence of global warming seek witnesses that cannot be tarnished by political muck and can serve as self-reporting instruments of global change. Glaciers—remote, pristine, detached—fit the bill. Through the visual record generated in the nineteenth and twentieth centuries, repeat photographs seem to speak for themselves and also for global warming. But in precisely those characteristics, which appeared to place them above and beyond the all-too-human, lay seeds of contention.

Glaciologists sought straightforward, unobjectionable evidence. What they got was a complicated type of visual representation with a checkered history. The iconography of ice came with undesirable consequences that were rooted in the history of repeat glacier photography. Some of these the glaciologists foresaw, some they did not. Repeat photographs oversimplified crucial scientific details about glaciers. They also reduced the problem of global warming to one of a distant, unpeopled wilderness, a misrepresentation that is both incorrect and harmful. The limits of repeat glacier photographs as icons of global warming were scientific and political.

Scientific Limits

The return to repeat glacier photography as evidence of global warming was grounded in the idea that seeing is believing. But is it, really? Qualitative representations like repeat photographs cannot tell you *why* glaciers are receding. This was precisely the critique of geophysical glaciologists like Robert Sharp. And they were not wrong. When Tad Pfeffer picked up former vice president Al Gore's book, *An Inconvenient Truth*, in 2006, he was startled by a photograph of the Columbia Glacier's dramatic recession, presented as evidence of global warming. As had been predicted by Austin Post, between 1980 and 2005 the Columbia had retreated 10 km—a considerable distance involving enormous volumes of ice. Pfeffer had spent much of his career studying the Columbia's dynamics. He knew the reasons for its retreat better than anyone, and in 2006 he knew it was not a simple case of retreat in response to global warming. The long-standing research program begun by Meier and Post and carried on by Pfeffer had shown that tidewater glaciers advance and retreat due to interactions between internal ice dynamics and the topography of the glacier's bed *combined with* climatic factors. Climate was only one of the variables involved in the Columbia's rapid and drastic retreat—the shape of its bed, its connection to the sea bottom, and internal pressures and tensions all played pivotal roles. Six years later, scientists determined that the Columbia's retreat had indeed been triggered when global warming destabilized the position of the terminus, setting off a chain reaction of complex dynamics. In 2006 this was not known. To the best of Pfeffer's knowledge at that time, the glacier's retreat could not be simply attributed to global warming.

Pfeffer had been a consultant for Gore's team, explaining the difficulties involved in extracting an anthropogenic climate signal from the Columbia's recession. Yet, there it was in that immensely popular book, testifying as evidence of global warming, exemplifying and reinforcing the equation: melting glaciers = global warming. But in this case, repeat glacier photography hid complex realities known to scientists. Pfeffer worried that when Al Gore, an "ally with a loud voice" but no technical training, used the Columbia as an example of human-induced glacier recession, "anybody who actually knows what is going on can challenge that easily and, you know, reduce his credibility—and of course this was done."[3] Indeed, people in the know—and people not in the know but ready to pounce on any misstep taken by climate activists like Gore—did just that. Gore's claims about glaciers have been gleefully attacked by scabrous online trolls and global warming denialists.[4] This potential for setting up "straw men" to knock down with bad arguments is precisely what scholars like Mark Carey caution is the danger of using the iconography of ice.[5] Indeed, one group of glaciologists noted in 2015, "images of retreating glaciers have become widely publicized illustrations of anthropogenic climate change [yet] the lagged response of glacier extents to climate change complicates the attribution of the observed changes to any particular cause."[6] In other words, we don't know enough about the physics of ice to make bold claims about the effects of climate based simply on pictures. This could easily have come from the pen of Robert Sharp. Geophysical glaciology's critiques of repeat photographs as ambiguous and not-very-reliable portrayals of glacier responses to climatic change still hold. Repeat glacier photographs do not simply speak for themselves.

Icons reduce complex phenomena into potent, charismatic visual representations. Their power is in association. Once an association is established, icons may stand for a feature of the world without accurately representing it. Consider red dragons: they are iconic images of Wales, but few people today expect to find one there. Generally, repeat photographs of retracting glacier tongues stand for global warming because they represent an important feature of it: glacier recession. But, as the foregoing has demonstrated, glacier dynamics are complex and many factors impinge on their ebb and flow. In some cases, global warming has caused glaciers to gain mass, though this is unlikely a long-term trend. None of this is captured in the simple visual association that melting glaciers equals global warming. Repeat glacier photographs as icons of global warming can be inaccurate and even misleading, as was the Gore team's use of the Columbia. Knowing what we know now about their bumptious history among those who study glaciers, we can appreciate Pfeffer's concerns, rooted in historically grounded values of the profession.

Cultural Limits

When glaciologists were casting about for public-facing forms of evidence, the history of glacier study made it more likely that they would gravitate toward one type of evidence—repeat photography—than others. There was already a large body of photographic work from glacier naturalists upon which to build. Those predecessors' photographs were more than just tourist snapshots (though they were sometimes that too): they were visual data with station locations and descriptions of how they were produced. Motivated by research agendas specific to their time, the photographs were nevertheless future-oriented, taken with the hope that someone would re-create them. It is small wonder they were taken up again. Given this precedent, repeat photographs seem a likely choice. Consider a counterfactual alternative. When glaciologists sought public-facing forms of evidence, acoustical recordings of calving events were a less likely option because there was little precedent for such work in the previous history of glacier study.

There was no body of prior work to rely on for comparison. What came before mattered for what came after. History matters.

Twentieth-century repeat glacier photographers were like archivists: they sought to create materials that would be useful in the future. Like archivists must, they did so with imperfect knowledge of what the future may need, armed only with their contemporary notions about what matters. Choices about what are significant today, made in real-life scenarios of finite resources, social priorities, and political pressures, shape and constrain what future historians will be able to know about the past. Similarly, glacier naturalists shaped what was possible for future iterations of repeat glacier photography.

Glacier naturalists were mostly confined to glacier termini. This was the area that piqued their curiosity and was most accessible, so the photographs they took focused on the end points of glaciers. They sought to understand glaciers as natural objects retreating from ice ages, and portrayed the ice alone in the frame. The emphasis on imposing glacier fronts and the dearth of human traces slotted these images snugly into a wilderness aesthetic. They recalled the sublime and the magisterial wild. All photographs shape how we see things, but the specifics of how repeat glacier photographs winnowed understandings of glaciers and global warming were rooted in the history of glacier study. They were rooted in how glacier naturalists perceived the landscape. The Rockies were Mary Vaux's wilderness retreat from Philadelphia's high society. Harry Fielding Reid, Ralph Tarr, and Bill Field, following John Muir, pictured the glaciers of coastal Alaska as a land reborn, possessing only geological history. Re-creating their photographs, frame for frame, nearly one hundred years later, reinforced ideas about glaciers as wilderness landscapes.

The highly wrought wilderness aesthetic of repeat glacier photographs has inspired some, like myself, to care about global warming for the sake of these beautiful places. It also placed glacier photographs within a familiar Western visual tradition that came with a built-in politics of nature's authority. But, as environmental historians have been saying since the 1990s, there are problems with "wilderness." Wilderness offers a limited picture of nature, one of distant, uninhabited landscapes as a salve for corrupted modernity. "True" nature, says wilderness, is found in faraway, seemingly empty lands, untouched by human activity. A preoccupation with wilderness, historians caution, diverts our attention from mundane nature and devalues the mixed landscapes in which most of us abide.[7] While it may inspire concern, it is less successful when it comes to spurring action. Repeat glacier photographs portray glaciers as tragic victims in a global crisis. These are declensionist narratives, where decline is a forgone conclusion.[8] They fail to forge intellectual, emotional, and practical links between concern about global warming and how we act in the places we live—and also how we experience global warming. While it is unreasonable to ask glaciologists to represent all aspects of global warming, it is important to recognize the limits of their representations. Repeat glacier photographs do not tell you how large-scale forces that frame how we live our lives, like global capitalism, commodity markets, infrastructures, and the consumer practices of a throw-away society, shape global warming and constrain what we can do about it. They don't tell you the causes or who is to blame. They also don't say much about the human consequences and how to manage them.

Representations of wilderness are also often inaccurate and contribute to long-standing misrepresentations of land. In North America, seemingly empty and sublime wildernesses are always someone's ancestral or present-day homeland. Picturing them as uninhabited suggests that people were not or are not there, which contributes to the erasure of Indigenous presence, past and present. If global warming, like pollution, is yet another twist in long histories of colonialism, depicting it as a problem for wildernesses that are actually someone's homelands furthers the displacement and structural erasure it relied upon to begin with.[9]

Repeat glacier photographs are doubly problematic in this regard—in what they depict and how they were made. Their production historically relied upon and contributed to the displacement of Indigenous people. The Vauxes' photographs helped displace and erase the Îyahê Nakoda (Stoney Nakoda), Tsuut'ina, Kainai, Pikuni, Niitsitapiksi, and Métis Peoples from what became Canada's Rocky Mountain national parks. Photographs taken by Field, Post, and others, following in the tradition of Muir's Alaska photographs, portrayed the southeastern corner of the state as a final frontier for explorers, scientists, and settlers, not the home of the Tlingit, Dené, Eyak, Sugpiaq, Aluutiq, and others for generations uncounted. Today's repeat photographs are direct descendants of these photographs. Through their format and lineage, problematic cultural assumptions of their forebears have been smuggled into current environmental discourses.

It would be unfair to place all of this on the shoulders of glaciologists who were simply using the tools at their disposal to address a pressing problem they encountered. At the turn of the twenty-first century it seemed many nonscientists did not believe global warming was a real threat even though photographs of receding ice seemed to say unequivocally that it was. Certainly, glaciologists did not intend for their photographs to have negative consequences for how people perceive global warming and glaciarized lands. Yet the meaning of an image is not only to be found in the intentions of its maker, but also in the interpretations of beholders. The point is not to assign blame. Nor is it to suggest glaciologists ought to stop taking repeat photographs; they indeed serve useful documentary purposes. Rather, the point is to stress that visuals, even of ostensibly bare ice, have histories and politics. Repeat photographs may have helped counter the politics of global warming denial, but they were enmeshed in historically generated politics of nature, land, home, and colonialism.

Glacier photographs do not speak for themselves; even within a relatively short span, repeat glacier photographs have meant different things. They have served as evidence of ice ages, glacier physics, global warming, and sometimes not much at all; they have conveyed ideas about wilderness, nature, and global warming. Glaciers are stubbornly, unavoidably polyvocal. This is why we need many hands, eyes, ears, voices, and minds working toward understanding glaciers and global warming; this is why we need humanists. Global warming and the ways it gets represented are not simply problems of nature; they are also problems of history and of politics.[10] Ice, Sverker Sörlin has observed, has become historical.[11] Something as seemingly straightforward as repeat glacier photography has been entangled with other historical processes in sometimes surprising, looping ways. What they meant to the photographers, and how they were produced and used, were imbricated with shifting ideas of scientific evidence, colonial development of parks and wilderness areas, geopolitical anxieties, and the politics of industry-backed science denial. This multistranded story highlights the strengths and weaknesses of glacier photographs: under what conditions they can be considered persuasive and where their limits lie. Such contours only become visible when we think historically about representations of global warming.

Repeat glacier photographs oversimplify the problem by making it seem that there is only one way of picturing glaciers: as vanishing wilderness areas facing a global catastrophe, knowable by natural science alone. Yet the winding history of their production allows us to better understand why they have the kind of meaning they do. This history shows that part of the problem is rooted in their making but also that resources for thoughtful, critical engagement with them can be found within that same history. Far from reinforcing simplistic dichotomies of pros and cons or believers versus skeptics, this history shows us that there are many ways to understand repeat glacier photographs, and that (for sometimes good reason) they haven't always been valued as a form of evidence for learning about glaciers. It helps us appreciate the contingency of picturing glaciers and global warming this way. Perhaps it could be otherwise?

Otherwise

There are as many ways of representing global warming as there are manifestations of it. Scientists, humanists, artists, and everyday activists are exploring multiple options to improve how we picture and think about global warming, too many to detail here.[12] Besides, this is a book about glaciers. Receding glaciers must be part of our conversation about global warming. Glaciers provide critical freshwater resources for many of us, and they support work of all kinds, from the labor of engineers at hydroelectric plants to ice cave tour guides. They also sustain life for plants and other critters. And they are themselves lively. Scientists are only just beginning to understand them as ecosystems, home to microbes, insects, and algae.[13] None of this is visible from repeat photographs of wild, isolated glacier fronts. Yet, repeat glacier photography need not be a matter of wilderness. Glaciarized lands have been peopled throughout the history of photography, and there are precedents for re-creating scenes that capture ongoing and changing entanglements between humans and ice. In the 1930s Martín Chambi (Jiménez), one of Peru's first Indigenous photographers, captured scenes of the Andean festival of Qoyllur Rit'i (Snow Star). This is a syncretic ritual merging pre-Columbian and Catholic beliefs in a high-altitude search for miracles and sanctity. Participants sing and dance their way up the Sinakara Valley in the Colquepunko Mountains. A select group of masked male dancers called ukukus ("tricksters of the glacier," according to American Peruvian artist Vincente Rivella) retrieve pieces of sacred ice from glaciers at the valley's head. Ukukus must be cunning to avoid falling into crevasses or being taken by one of the condenados, the evil spirits condemned, like Sisyphus, to perpetually carry boulders of ice up the glacier. The "Lord's Ice"

FIGURE 25 *Peregrino en Qoillur Rit'i, Oncongate*, 1934. Photo by Martín Chambi.
Courtesy of the Martín Chambi Photographic Archive, Cusco–Peru. 2018.

FIGURE 26 *La Cordillera Colquepunku, Peru*, 2004. Photo
by Eirik Johnson. Courtesy of Eirik Johnson.

(it belongs to both Christ and the Lord of Qoyllur Rit'i) is then brought down to the city of Cusco and distributed among the faithful in time to coincide with the feast of Corpus Christi.[14] In the ritual of Qoyllur Rit'i, ice is quite literally transmogrified into an icon.

Chambi's photograph *Peregrino en Qoyllur Rit'i* (*Pilgrim at Qoyllur Rit'i*) (1930s) (fig. 25) captures a man in a woven poncho and cap sitting on a promontory above the crowds and campfires below, gazing contemplatively into the distance. Over seventy years later New York–based artist Eirik Johnson took a photograph echoing Chambi's (fig. 26). Johnson's photograph is framed from a similar perspective as Chambi's, looking up the valley toward the ice from a high vantage, with both pilgrims and glaciers appearing further away. The celebrants, tiny figures against a dry montane landscape, appear ant-like. Lacking the contemplative human subject, and with the glaciers clearly withdrawn, Johnson's photograph elicits nostalgia and loss. Yet it is not a lament for lost wilderness. The festivalgoers are still there, as are the glaciers, though diminished. The photograph thus evokes resilience and continuity in the face of overwhelming odds, placing people adamantly among their glaciers, even as the latter recede. The combination of change and obdurate constancy suggests an element of solastalgia—a grieving for home transformed.[15] Certainly, this repeat pair goes beyond using photographs to show *that* global warming is happening, reinforcing the dichotomy of us-them camps of believers and skeptics. They show a specific way in which ice is entangled in people's lives and what the lessening of that ice might mean to them. They demonstrate that glaciers are natural and cultural, and we need the contributions of many types of investigators to help us see the complex ways our lives are tied, unevenly to the yet-undisclosed fates of mountain glaciers.

When I first met the Bow Glacier in 2003, I could not have predicted that our meeting would weave into a meandering (sometimes tortuous) decade of researching and writing about glacier representations. What I experienced then as a liberating wilderness now appears as a place of deep history, continual change, and diverse human and other-than-human entanglements. A place where glaciers have said many different things to many different folks. It has made me care for them as crossroads and archives for an array of perspectives and experiences. Now when I see a portrait of ice I wonder: What did the photographer see when peering through the lens? What did the glacier see?

NOTES

Introduction

1. Indeed, scholars have explored ice as a medium of memory for just these reasons. See Frank and Jakobsen, *Arctic Archives*.

2. In seeking such stories this book joins the work of scholars like M Jackson, Mark Carey, Karine Gagné, and Sverker Sörlin, who have criticized and sought to go beyond what I call the iconography of ice. My debt to their work will become apparent.

3. A note on terminology: scientists, politicians, and scholars use a variety of phrases to refer to the idea of climatic change induced by human action (e.g., "greenhouse warming," "anthropogenic climate change," "global climate change," "global weirding," "CO_2-induced climate change," and so forth). Following Joshua P. Howe and Mike Hulme, I use "global warming" to refer to "human-caused climate change, plus politics." See Howe, *Behind the Curve*, 13–14; and Hulme, *Why We Disagree about Climate Change*, xxxviii–xxxix.

4. Barasch, *Icon*, 2.

5. Jackson, *The Secret Lives of Glaciers*, 9.

6. Jackson, *The Secret Lives of Glaciers*; Jackson, "Glaciers and Climate Change; Carey, "The History of Ice"; Garrard and Carey, "Beyond Images of Melting Ice"; and Carey, "The Trouble with Climate Change."

7. For a sampling of the foundational historical literature on wilderness as anti-modernity, see Cronon, "The Trouble with Wilderness"; Oelschlaeger, *The Idea of Wilderness*; and Nash, *Wilderness and the American Mind*. On the evocative power of mountain landscapes as wildernesses, see Cosgrove and della Dora, *High Places*, 1–16; Macfarlane, *Mountains of the Mind*; Schama, *Landscape and Memory*; and Nicholson, *Mountain Gloom and Mountain Glory*.

8. S. A. Inkpen, "Are Humans Disturbing Conditions in Ecology?"

9. S. A. Inkpen, "Demarcating Nature, Defining Ecology."

10. Gina Rumore makes this link explicit in her study of succession ecology and glaciology at Glacier Bay National Monument. See Rumore, "A Natural Laboratory, a National Monument."

11. Cronon, "Trouble with Wilderness"; Cosgrove, "Images and Imagination; Carey, "History of Ice."

12. Dunaway, *Seeing Green*, 2.

13. This is an area of scholarly debate. Andreas Malm and Alf Hornborg, in "The Geology of Mankind?," have emphasized the ability for the global rich to shelter from the effects of global warming, while Dipesh Chakrabarty, in "The Climate of History," has underlined the negative existential threat that it poses to all humans if left unmitigated.

14. Heglar, "Climate Change Ain't the First"; Hecht, "The African Anthropocene"; Whyte, "Indigenous Experience"; Whyte, "Is It Colonial Déjà Vu?"; Bennett et al., "Indigenous Peoples, Lands, and Resources." Similar points have been made regarding pollution in Liboiron, *Pollution Is Colonialism*.

15. Whyte, "Indigenous Experience," 158.

16. Ford et al., "The Resilience of Indigenous Peoples."

17. Jackson, "Glaciers and Climate Change."

18. Garrard and Carey, "Beyond Images of Melting Ice," 102.

19. Carey, "The Trouble with Climate Change," 259.

20. Artists, scientists, and scholars have become much more creative in depicting global warming, yet blue glacier ice remains an important part of the visual repertoire. See, for instance, Schmidt and Wolfe, *Climate Change*; Fowkes and Fowkes, *Art and Climate Change*; and Sheppard, *Visualizing Climate Change*.

21. Jackson, *The Secret Lives of Glaciers*; Gagné, *Caring for Glaciers*.

22. Carey, *In the Shadow of Melting Glaciers*; Taillant, *Glaciers*; Cruikshank, *Do Glaciers Listen?*; Carey et al., "Glaciers, Gender, and Science"; Orlove, Wiegandt, and Luckman, *Darkening Peaks*; Sörlin, "Can Glaciers Speak?"; D. Inkpen, "Ever Higher."

23. Exceptions to this claim are Garrard and Carey, "Melting Ice"; Sörlin, "The Global Warming that Did Not Happen"; and D. Inkpen, "Of Ice and Men."

24. These ideas about critique and history as a narrative art that both explains past phenomena and emphasizes the contingency of the past with a liberatory potential are indebted to both Michel Foucault's interpretation of the critical tradition in Western philosophy and Dipesh Chakrabarty's ideas about history's capacity to provincialize dominant histories. More personally and immediately, my thinking in these ways has been molded by the writings and teaching of philosopher and historian John Beatty and conversations with my philosopher-historian husband, Drew Inkpen. For published sources see Foucault, "What Is Critique?"; Foucault, "Introduction to Kant's *Anthropology*"; Chakrabarty, *Provincializing Europe*; and Beatty, "Narrative Possibility and Narrative Explanation."

25. *Life* magazine, February 2, 1962, R1–R2.

26. Silverman, *The Miracle of Analogy*, 140. Silverman makes this remark while con-

sidering Benjamin's idea of the optical unconscious in two of his essays: "The Work of Art in the Age of Mechanical Reproduction" (1935) and "A Little History of Photography" (1931). I have found Silverman's book to be a productive source to think with.

27. This definition is taken from National Snow and Ice Data Center, "All About Glaciers," accessed December 11, 2017, https://nsidc.org/cryosphere/glaciers/questions /what.html. For an accessible exposition of glacier features and processes see Taillant, *Glaciers*, 24–49.

28. Cogley et al., "Glossary of Glacier Mass Balance."

29. World Glacier Monitoring Service, "Latest Glacier Mass Balance Data," accessed September 23, 2021, https://wgms.ch/latest-glacier-mass-balance-data/.

30. Sanford, *Our Vanishing Glaciers*, 59; M. Demuth, *Becoming Water*, 37.

31. D. Inkpen, "Ever Higher"; Knight, *Glaciers*.

32. Leading research in this area is being conducted by the Columbia River Inter-Tribal Fish Commission. See Dittmer, "Changing Streamflow," 638.

33. Cruikshank, "Are Glaciers 'Good to Think With'?"

34. Cruikshank, *Do Glaciers Listen?*; Gagné, *Caring for Glaciers*.

35. Jackson, *The Secret Lives of Glaciers*.

36. Carey, *In the Shadow of Melting Glaciers*, 25, 191–93; Qiu, "Ice on the Run."

37. This is the role allotted to glacier study in Fleming, *Historical Perspectives on Climate Change*; Weart, *The Discovery of Global Warming*; and Edwards, *A Vast Machine*.

38. According to Sverker Sörlin, glaciology illustrates discontinuity and fragmentation in the history of climate science. Glaciologists, he contends, were latecomers to the scientific consensus on climate change, which suggests the existence of alternatives to the dominant histories of climate science. Certainly, his insight about contingency and discontinuity are valid. However, Sörlin's thesis is built on the work and legacy of Swedish geographer and glaciologist Hans W. Ahlmann, whose theory of polar warming excluded the possibility that humans could be the cause. Of course Ahlmann's work and legacy do not exhaust glaciological contributions to global warming research. Sörlin, "The Anxieties of a Science Diplomat" and "The Global Warming that Did Not Happen."

39. Webb, Turner, and Boyer, "Introduction."

40. Silverman, *The Miracle of Analogy*, 81–83.

41. Webb, Turner, and Boyer, "Introduction," 3.

42. The literature on visualizations in science is vast; a few seminal studies include: Tucker, *Nature Exposed*; Daston and Galison, *Objectivity*; Schiebinger, *Nature's Body*; Latour, "Visualization and Cognition"; Jordanova, *Sexual Visions*; Haraway, *Primate Visions*; and Rudwick, "The Emergence of a Visual Language."

43. Stepan, *Picturing Tropical Nature*, 14.

44. Mark Carey argues that media portrayals of glaciers in climate change discourse say as much about ideas of parks, wilderness, and people-nature relations as they do about climate change. Carey, "The Trouble with Climate Change," 259.

45. Those familiar with the history of mountaineering and landscape photography might wonder why Boston mountaineer and geographer Bradford Washburn does not feature prominently in this story. Washburn is one of the most celebrated mountain photographers of the twentieth century and took many photographs of glaciers. The reason for Washburn's absence, and those of other mountain photographers who might come to mind, is that I have attempted to follow those whose photographic work was in the service of *scientific glacier study*. Washburn was predominantly interested in mapping. Michael Sfraga has written an engaging and informative biography of Washburn: *Bradford Washburn: A Life of Exploration*.

46. Some of what Lorraine Daston has termed our "disciplinary dizziness," over what, exactly, the history of science is the history of can be explored in a collection of essays: Kohler and Olesko, *Clio Meets Science*; and Daston, "The History of Science."

47. The scientific regimes I describe here are similar to the "codes of epistemic virtue" invoked by Daston and Galison in their history of objectivity: truth-to-nature; mechanical objectivity; and trained judgment. Although Daston and Galison demur from the term "regime" because it seems to invoke a succession of political regimes, to me this hesitancy seems misplaced, as successive political regimes do not in fact entirely replace their predecessors. Tattered trappings of bureaucracy always remain. Daston and Galison, *Objectivity*, 18–19.

48. See, for instance, Etienne Benson's recent treatment of the history of environments and environmentalisms in terms of "environments of empire," "the biosphere as battlefield," and "the human planet." Benson, *Surroundings*.

49. Throughout the three regimes covered, scientific life in the field intersected with mountaineering practice. The social and cultural worlds of North American mountaineering shaped who did glacier research and how they did it. The gendered and colonial legacies of early mountain recreation infused glacier photography, and suggest ways in which the practices and structures of natural history remain alive today. Even today a background in mountaineering can be advantageous for work in field glaciology. See Savage, "Climatology on Thin Ice." Historians of science have written about the coupling of recreation and the field sciences in the early twentieth century. See Kohler, *All Creatures*; and Kohler, "History of Field Science." Some have paid particular attention to the gendered politics of recreation and field science. See Hevly, "The Heroic Science of Glacial Motion"; Reidy, "John Tyndall's Vertical Physics"; Reidy, "Evolutionary Naturalism on High"; and Reidy, "Mountaineering, Masculinity."

1. Documenting

1. The Vaux family folders contains 2,924 photographs, 2,632 negatives, 167 transparencies, and 125 prints. The Vauxes have been the subject of several biographical studies: Edward Cavell's *Legacy in Ice: The Vaux Family and the Canadian Alps*; Henry Vaux Jr.'s *Legacy in Time: Three Generations of Mountain Photography in the Canadian West*; and Marjorie G. Jones's *Life and Times of Mary Vaux Walcott*.

2. Robinson and Slemon, "The Shining Mountains," 116.

3. Snow, *These Mountains Are Our Sacred Places*, 19.

4. Putnam, *The Great Glacier and Its House*.

5. Endersby, *Imperial Nature*; Camerini, "Wallace in the Field."

6. The historical literature on natural history is extensive; some of the canonical studies include Coleman, *Biology in the Nineteenth Century*; Jardine, Secord, and Spary, *Cultures of Natural History*; Farber, *Finding Order in Nature*; Browne, "A Science of Empire"; Kohler, *All Creatures*; and Foucault, *The Order of Things*.

7. Imbrie and Imbrie, *Ice Ages*, 19–31.

8. They followed his stricture until his daughter, Mary (at seventy-five), traveled to Japan. See Jones, *Life and Times of Mary Vaux Walcott*, 145.

9. "Guarded" refers to the idea that a traditional Quaker education protected children from the influences of the broader society.

10. Nineteenth-century American Quakers adopted a variety of attitudes toward the natural sciences. It is thus imprudent to seek explanations for the Vauxes' scientific interests in their Quakerism. It is better, as many scholars have argued, to see the Vauxes' scientific pursuits, like their travel and mountaineering, as part of Victorian bourgeoise life. Cantor, "Quakers and Science"; Hindle, "The Quaker Background and Science"; Stanley, "'An Expedition to Heal the Wounds of War.'"

11. Mary and George were briefly associated with the avant garde photo-secession movement headed by Alfred Stieglitz, but they steered clear of club debates about the relation between photography and art. See Cavell, *Legacy in Ice*, 17.

12. Hansen, "Albert Smith," 137. See also Willen, "Composing Mountaineering."

13. William S. Vaux, "Taking the Glaciers," *Minneapolis Journal*, September 3, 1898.

14. Robinson, *Conrad Kain*, xxxi.

15. Mary met Walcott (1850–1927) near Field, BC, where he was working on the famed Burgess Shale. In 1914 they married and she moved to Washington, DC, where she became involved in establishing a Society of Friends for US president and fellow Quaker Herbert Hoover and his wife, Lou. Jones, *Life and Times of Mary Vaux Walcott*, 69–80.

16. Jones, *Life and Times of Mary Vaux Walcott*, 135.

17. Darling, "Up the Bow and Down the Yoho," 158.

18. Mary Vaux to Charles Walcott, March 11, 1912, quoted in Skidmore, *This Wild Spirit*, 210.

19. Cavell, *Legacy in Ice*, 13.

20. Mary Vaux to Charles Walcott, February 19, 1912, quoted in Skidmore, *This Wild Spirit*, 204.

21. Quoted in Cavell, *Legacy in Ice*, 12.

22. Diary entry, William S. Vaux Jr., July 17, 1887, folder 1, series 5, Vaux Fonds.

23. Jones, *Life and Times of Mary Vaux Walcott*, 41–42.

24. Vaux and Vaux, "Some Observations," 122.

25. Champoux and Ommanney, "Evolution of the Illecillewaet Glacier."

26. Scott, *Pushing the Limits*, 40–41.

27. Green, *Among the Selkirk Glaciers*, 67.

28. Jackson, *Secret Lives of Glaciers*, 61–63. Jackson draws on the recovery efforts by glaciologists Richard S. Williams and Oddur Sigurðsson as described in their introduction in Pálsson, *Draft of a Physical, Geographical, and Historical Description*, xviii–xxxvi; see also Þórarinsson, "Glaciological Knowledge."

29. Ladurie, *Times of Feast, Times of Famine*.

30. Agassiz, *Études sur les glaciers*; Forbes, *Travels through the Alps of Savoy*. See also Shairp, Tait, and Adams-Reilly, *Life and Letters of James David Forbes*, 329–30.

31. Green, *Among the Selkirk Glaciers*, 218–19.

32. Albrecht Penck to William Vaux Jr., July 5, 1899, folder I.B.I.a, series 5, Vaux Fonds.

33. Scott, *Pushing the Limits*, 42.

34. Harold Topham to William Vaux Jr., July 7, 1899, series I.B.I.a., Vaux Fonds. Topham's 1891 article merely mentioned that he left marks on boulders and that Illecillewaet had retreated since Green's 1888 visit. Topham, "The Selkirk Range, North-West America," 554.

35. Vaux and Vaux, "Additional Observations," 508.

36. The naming of both "Photographers' Rock" and Rock E suggests that photographic and survey stations acquired meaning through more of a mundane process than the one described by D. Graham Burnett in his account of the Transverse Survey of Guyana. Rock W was, like the stations of the Transverse Survey, a natural-artificial construction born of the needs of surveying. However, its significance was less lofty than the historical, mythical, and colonial import bestowed upon Schomburgk's stations or through what Burnett describes as a process of *consummatio*. Burnett, *Masters of All They Surveyed*, 133–45.

37. Vaux and Vaux, "Observations Made in 1906," 569.

38. Tyndall, *The Forms of Water*, 163–91. For the reach of the dispute see Ball, "On the Cause of the Descent of Glaciers"; Croll, "On the Physical Cause of the Motion of Glaciers"; Moseley, "On the Descent of Glaciers"; Matthews, "Mechanical Properties of Ice"; and Wallace, "The Theory of Glacier Motion."

39. Hevly, "Heroic Science of Glacial Motion"; Reidy, "John Tyndall's Vertical Physics"; Reidy, "Mountaineering, Masculinity"; Oreskes, "Objectivity or Heroism?"

40. Vaux and Vaux, "Observations Made in 1906," 569.

41. Twelve plates were made by CPR employees at Revelstoke. Nine were placed; only eight were used. Each plate measured six inches square and one-eighth inch thick, was given three coats of red paint, and then was labeled VAUX No. 1–8 1899. The paint—an oil with a flake-graphite base—was made especially for the plates by H. E. & D. S. Yarrall; it came in small cans fitted with a special spring top to make it easily packed, carried, and opened and closed while in the field. The plates had to be replaced in 1906 and 1909 because many had been swallowed by the glacier over the years. See

Glacial Notes, August 1910, folder I. B. 2, series 7; "George Vaux," folder I.C.2.a.iii, series 7; folder 2, I. B.I b i, series 5, Vaux Fonds.

42. Vaux and Vaux, "Observations Made in 1900," 215.

43. Bailey, *Leisure and Class in Victorian Britain*.

44. Vaux and Vaux, "Some Observations," 121.

45. Francis, *National Dreams*.

46. Ham, *Reminiscences of a Raconteur*, 59962.

47. MacBeth, *The Romance of the Canadian Pacific Railway*; Lamb, *History of the Canadian Pacific Railway*; Cruise and Griffiths, *Lords of the Line*.

48. For comparison with the American transcontinentals, see White, *Railroaded*.

49. Bliss, *Right Honorable Men*, 22.

50. Francis, *National Dreams*, 23; Brglez, "Surveying Indigenous Space."

51. This assessment of the nation's "fathers" is from Earle, "Three Stories, Two Visions," 346.

52. Francis, *National Dreams*, 14.

53. Otter, *The Philosophy of Railways*, 15.

54. For a partial but extensive list of Canadian songs about railways and trains for your next transcontinental playlist, see "Canadian Railway Songs," James R. Hay, accessed April 1, 2016, http://www.railwaysongs.ca.

55. Snow, *These Mountains Are Our Sacred Places*, 106; see also Hart, *The Place of Bows*, 114.

56. Binnema Niemi, "'Let the Line Be Drawn Now,'" 728.

57. Bella, *Parks for Profit*, 20.

58. See Jacoby, *Crimes Against Nature*; and Spence, *Dispossessing the Wilderness*.

59. This process was not achieved overnight and not until 1893 was hunting banned within park boundaries. See Snow, *These Mountains Are Our Sacred Places*, 80. See also Mason, "The Construction of Banff"; Robinson and Slemon, "Shining Mountains"; and Loo, *States of Nature*, 39–62.

60. The first Swiss guide to appear in the Rockies was Peter Sarbach, who was hired by the Appalachian Mountain Club in 1897 following the first fatal climbing accident in the Rockies the previous year. Until 1954 the Canadian Pacific hired European guides to lead railway clients and teach climbing techniques. Several histories treat the making of the park as a climber's space through the labor of foreign guides: Robinson, *Conrad Kain*; Reichwein, *Climber's Paradise*; Costa and Scardellato, *Lawrence Grassi*; and Spaar, *Swiss Guides*.

61. Pringle, "William Cornelius van Horne," 79.

62. Taylor, *Pilgrims of the Vertical*; Smart, *Paul Preuss*.

63. Donnelly, "The Invention of Tradition"; Robinson, "Storming the Heights"; Reichwein, *Climber's Paradise*, 92–96.

64. The phrase "50 Switzerlands in One!" was coined by British mountaineer Edward Whymper while visiting the Rockies. He was, however, not the only one to

draw such comparisons. See Marsh, "The Rocky and Selkirk Mountains"; and Williams, "'That Boundless Ocean of Mountains,'" 76.

65. Other Canadian Pacific promotions that highlighted the mountaineering potential of the Rockies at this time were "Mountaineering in the Canadian Rockies" (1901) and "The Challenge of the Mountains" (1905).

66. Folder I.B.I.d.ii, series 7, Vaux Fonds.

67. Cavell, *Legacy in Ice*, 63.

68. Fay, "The Resort of Glacier House," 275.

69. It is perhaps a sign of the Canadian Pacific's valuation of its employees that in 1898 and 1899 it provided the services of guide Dan Fraser free of charge but asked the Vauxes to pay $1.50 for the pack pony.

70. Folder 4, series "Correspondence," subseries "Geographical Information and Inquiry," Vaux Fonds.

71. Vaux series V, "Correspondence, 1897–1915" and "1889–1902," WMCR.

72. Wilcox to William Vaux Jr., July 25, 1900, folder I.B.I.a, series 5, Vaux Fonds; Vaux and Vaux, "Additional Observations," 503.

73. In 1895 a list of observations that mountaineers could make that would contribute to glacier study appeared in the *Journal of Geology*. It called for: a general description; a state of fluctuation (advance, retreat, stagnancy); thickness of the ice; velocity of flow; and rates of melting. See Reid, "The Variations of Glaciers, XII," 285–88.

74. See Matthes, "Glacier Measurements in the United States," (draft), in folder 2, box 76, series 7, Field Papers; Field, "Glacier Studies in Alaska"; and Flint, "The American Alpine Club."

75. Parker, "The Alpine Club of Canada."

76. For biographical details on Wheeler, see Reichwein, *Climber's Paradise*, 17–19.

77. First developed at length by French colonel Aimé Laussedat, phototopographical (or photogrammatic) surveying solved the problem of mountainous terrain faced by the Dominion Topographical Survey. Until the Rocky Mountains, that team had used chains to measure baselines from which they plotted the land into six-square-mile townships. This worked for measuring flat ground, but mountainous terrain (and its potential for resource extraction rather than standardized parcels of land) made such methods unfeasible. Photogrammatic surveys, though they required more post-field labor, were the cheapest way to tackle the measuring of mountains. See Bridgland, "Photographic Surveying in Canada"; and MacLaren, Higgs, and Zezulka-Mailloux, *Mapper of Mountains*.

78. Quoted in Reichwein, *Climber's Paradise*, 15.

79. Elizabeth Parker, "The Canadian Rockies: A Joy to Mountaineers," *Manitoba Free Press*, September 23, 1905, 20.

80. See Hansen, *The Summits of Modern Man*, 245–74.

81. Reichwein, *Climber's Paradise*, 25.

82. Parker, "The Alpine Club of Canada," 7.

83. Donnelly, "The Invention of Tradition," 238–39.

84. Williams, "'That Boundless Ocean of Mountains,'" 75; Wheeler, "Affiliation with 'The Alpine Club.'" See also Robinson, "Storming the Heights."

85. Quoted in Reichwein, *Climber's Paradise*, 49; see also 45–48.

86. Parker, "The Alpine Club of Canada," 3; emphasis added.

87. He first met the Vaux family earlier that summer, on July 29. See Wheeler diary 1902–1903, folder M546 09–10, Wheeler Fonds.

88. Yeigh, "Canada's First Alpine Club Camp"; Wheeler, "The Canadian Rockies." From 1907 to 1928 the journal featured a "scientific section": of the sixty-four articles published there between 1907 and 1922, twenty-one concerned glacier studies, and of those, thirteen were written by Wheeler; six by the Vauxes. Other authors on glaciers included William Sherzer and the University of Toronto geologist Arthur P. Coleman, who contributed geological papers. The Scientific Section reappeared in 1932, then disappeared again until 1953, with William Henry Mathews as editor, who served until 1962. Its final iteration was edited by Wheeler's grandson John Oliver Wheeler from 1963 until 1970.

89. William O. Field, who was not affiliated with a particular section, was invited to participate at the suggestion of American alpinist Henry S. Hall Jr., but it does not appear that Bill Field accepted the invitation. Wharton Tweedy to Alexander MacCoubrey, September 17, 1932, folder AC 00M 010, ACC Fonds, WMCR. The following year the section gained H. W. Allan, H. E. Samson, Cyril Wates, and Kingman. MacCoubrey to Tweedy, January 6, 1933, folder AC 00M 15A, ACC Fonds. Wheeler to J. Monroe Thorington, January 27, 1932; memo, Wheeler to Thorington, November 28, 1932; and memo, Wheeler to Thorington, February 21, 1933, all in folder AC M106 58, Thorington papers.

90. Although he sought to employ mountaineers in the service of glacier study, Wheeler remained leery of amateurs, insisting that repeat photography was only useful if performed "systematically." See "Glacier Section," *Canadian Alpine Journal* 22 (1932): 21–22. See also Wates to MacCoubrey, August 11, 1936; Kingman to MacCoubrey, 9 September 1936, folder AC 0 47, ACC fonds.

91. Baird, "A Note on the Commission," 253. See also Radok, "The International Commission on Snow and Ice."

92. Initially ten nations were involved: Switzerland, Germany, Austria, Sweden, Norway, Denmark, Great Britain and Colonies, the United States, Russia, and France. Italy joined in the commission's second year. The first ten of the commission's annual reports were published in the *Archives des sciences physiques et naturelles de Geneve* and, after his election to presidency in 1906, by University of Vienna professor Eduard Brückner in his *Zeitschrift for Gletscherkunde*.

93. Thank you to Shawn Marchese and Alan Sisto for bringing this idea to my attention.

94. Vaux and Vaux, "Additional Observations," 510.

95. Brückner, "Fluctuations of Water Levels."

96. Forel, "Periodic Variations of Alpine Glaciers," 386.

97. Vaux and Vaux, "Additional Observations," 510; Vaux and Vaux, "Observations Made in 1906," 568.

98. Reid, "The Variations of Glaciers, XII," 47.

99. Albers and Bear, "Photography's Time Zones," 2. Joel Snyder argues that the capacity to picture the invisible is the specific role of photography. Snyder, "Res Ipsa Loquitur," 216.

100. Benjamin describes the optical unconscious as that which is revealed by the conscious penetration of space with the aid of the camera, which, in the process, exposes the normally undetected limits of human perception. Benjamin, "The Work of Art." For a metaphysical interrogation of the idea see Silverman, *The Miracle of Analogy*, 139–49.

101. See Reichwein, *Climber's Paradise*, 162–65; and MacCoubrey to the Executive Committee, July 27, 1936, folder AC 00 127, ACC Fonds.

2. Transitions

1. "General Alaska Water Facts," Alaska Department of Natural Resources, accessed January 12, 2018, http://dnr.alaska.gov/mlw/water/hydro/components/water-facts .cfm.

2. Arendt, "Assessing the Status of Alaska's Glaciers"; Colgan et al., "Monte Carlo Ice Flow Modeling."

3. Pfeffer, *The Opening of a New Landscape*, 2–3.

4. Post, "The Exceptional Advances of the Muldrow" and "Distribution of Surging Glaciers."

5. Environmental historians sometimes refer to such coalitions of nature and artifice as "envirotechnical systems." They are "historically and culturally specific configurations of intertwined 'ecological' and 'technological' systems." Pritchard, *Confluence*, 19. Fellow historian of the field sciences Jeremy Vetter uses the idea of envirotechnical system in a similar way by considering how the transportation technology of the railway, along with the telegraph, the postal service, and other "tools" broadly defined, "undergirded a new travel geography of the West that was as critical to the practice of science as to any other activity." Vetter, *Field Life*, 9.

6. The phrase used for the subhead is borrowed from Julie Cruikshank. It describes the orientation underlying her work to listen for the knowledge of glaciers held by Tlingit and Tagish women Annie Ned, Kitty Smith, and Angela Sidney in the stories they carried. This orientation was, she tells readers, prompted by an injunction from Annie Ned in 1982. While my purpose in this book is very different from Cruikshank's, I believe her orientation was valuable for anyone thinking about glaciers on Indigenous lands. See Cruikshank, *Do Glaciers Listen?*, 76.

7. Cruikshank, *Do Glaciers Listen?*, 29, 69.

8. Read of Amy Marvin in Dauenhauer and Dauenhauer, *Haa Shuká*, 275. See also Catton, *Inhabited Wilderness*, 10; and Cruikshank, *Do Glaciers Listen?*, 76–124, 158–60, for versions of the story with different details.

9. Cruikshank, "Good to Think With?," 244.

10. Archaeologists have reported evidence of a Little Ice Age settlement, precisely where Kaasteen's story places the Tcukanadi village prior to the onslaught of the glacier. Connor et al., "The Neoglacial Landscape."

11. Quoted in R. Campbell, *In Darkest Alaska*, 248.

12. Cruikshank, "Good to Think With?," 247.

13. William O. Field, August 19, 1950, Alaska 1950 Field Diary, folder 5, box 1, subseries a, series 1, Field Papers.

14. Roberts, "Walter Wood and the Legacies of Science."

15. B. Demuth, *Floating Coast*.

16. R. Campbell, *Darkest Alaska*, i.

17. "Kasteen," Glacier Bay National Park and Preserve, accessed June 16, 2022, https://www.nps.gov/articles/000/kaasteen.htm.

18. Menzies, *The Travel Diaries of Archibald Menzies*, 161–62.

19. "Kasteen," Glacier Bay National Park and Preserve.

20. Handy, "First Exploration of the Harvard Glacier."

21. Fox, *The American Conservation Movement*.

22. Worster, *A Passion for Nature*, 111–14, 120–47.

23. Muir was not even the first white person to explore the bay. Between Vancouver's travels and his 1879 arrival, the ice had retreated and the bay's open waters had been cloven by the prows of many ships, American and Russian, private and government-sponsored. Catton, *Inhabited Wilderness*, 12. Muir first published on his visit to Glacier Bay for the North Pacific Railroad and *American Geologist*. His "discovery" was made widely known by an expanded article in the 1895 *Century Magazine* piece, which included details from his 1880 trip. See Muir, *Alaska via Northern Pacific R.R.*, 3–17; Muir, "Alaska"; and Muir "The Discovery of Glacier Bay."

24. Muir, "Discovery of Glacier Bay," 234, 237.

25. Rumore, "A Natural Laboratory," 41. Foundational narratives of the park based in Muir's story—and reinscribed through pamphlets and signs quoting his writings—have recently been complicated by the return of the Tlingit to hunt, gather, and fish in Glacier Bay once again. See Mary Catharine Martin, "After Hundreds of Years, Huna Tlingit Return to Ancestral Home of Glacier Bay," *Juneau Empire*, August 31, 2016, http://juneauempire.com/art/2016-08-31/after-hundreds-years-huna-tlingit-return -ancestral-homeland-glacier-bay.

26. Goetzmann and Sloan, *Looking Far North*, 45.

27. See Schultz, "The Debate Over Multiple Glaciation"; and Numbers, "George Frederick Wright," 632–33.

28. Aalto, "Rock Stars."

29. Hubbard, Baker, and Johnson, "Memorandum of Instructions," 194.

30. Russell, "Climatic Changes Indicated by the Glaciers," 322.

31. Goetzmann and Sloan, *Looking Far North*, 5.

32. Ralph Tarr died in 1912, leaving his student Lawrence Martin, then a professor of physiography and geography at the University of Wisconsin, to complete their joint study in 1913. Tarr, *Alaskan Glacier Studies*; Tarr, "The Malaspina Glacier"; Martin, "Alaskan Glaciers in Relation to Life"; and Martin, "Glacial Scenery in Alaska."

33. Field and Brown, *With a Camera in My Hands*, 64.

34. Field and Brown, *With a Camera in My Hands*, 13.

35. F. Field, *From Right to Left*, 8–9.

36. W. Field, "In Search of Mount 'Clearwater,'" 9.

37. Field and Brown, *With a Camera in My Hands*, 27–29.

38. William O. Field to C. Hart Merriam, March, 23 (year unknown); and William O. Field to Edward S. Curtis, (n.d.); both in folder 1, box 4, series 12, Field Papers.

39. Field and Brown, *With a Camera in My Hands*, 28.

40. Field to Curtis, (n.d.), folder 1, series 12, box 4, Field Papers.

41. Field and Brown, *With a Camera in My Hands*, 35, 40.

42. Scott, *Pushing the Limits*, 99–101; Stevenson and Wright, "The Living Labyrinth," 58. They also hired the less remarkable Percy Pond of Juneau.

43. William O. Field, Alaska Diary, 1926, folder 1, box, 1, sub-series a, series 1, Field Papers.

44. W. Field, "The Fairweather Range," 465, 468–70; W. Field, "Active Change Noted in Alaskan Glaciers," *Crimson*, October 18, 1926, folder 4, box 17, sub-series b, series 1; Field and Brown, *With a Camera in My Hands*, 35.

45. Field and Brown, *With a Camera in My Hands*, 51–63.

46. Field and Brown, *With a Camera in My Hands*, 5.

47. Tim Malchoff and Kari Brookover, "Traditional Place Names," Chugachmiut Heritage Preservation, accessed April 13, 2022, https://chugachheritageak.org/traditional-place-names.

48. Gilbert, *Glaciers and Glaciation*, 71–97; Grant and Higgins, "Coastal Glaciers of Prince William Sound," 7–40; Tarr, *Alaskan Glacier Studies*, 232–388. The USGS sent surveyors in 1905, 1908, 1909, 1913, 1914, 1925, and 1931; Field did not know of the 1931 survey until after he returned.

49. William O. Field, Prince William Sound Diary, August 29, 1931, folder 2, box 1, sub-series a, series 1, Field Papers.

50. William O. Field, Alaska 1931 trip summary, folder 2, box 18, sub-series b, series 1, Field Papers; W. Field, "The Glaciers of the Northern Part of Prince William Sound," 373, 376.

51. It may seem odd that a botanist was interested in glacier recession, but Cooper saw it as an opportunity to study succession, a theory then dominating botanical ecology. The repopulation of flora on grounds recently uncovered by retreating ice

presented an ideal experiment location. In turn, glacier studies benefited from botanical ecology. Succession research provides estimates of when the ice had receded in the past based on the age and composition of the flora surrounding it. As carbon-dating technology became more refined, the remnants of ancient trees further improved the accuracy of these assessments. Field later worked with Cooper's student Donald Lawrence and Calvin Heusser, the latter a botanist from Oregon State University. On Cooper's role in the history of Glacier Bay and succession ecology, see Rumore, "A Natural Laboratory."

52. Cooper's "The Problem of Glacier Bay" definitively laid out the geological history of the bay. Cooper, "The Problem of Glacier Bay, Alaska"; W. Field, "Observations on Alaskan Coastal Glaciers."

53. Wright, *Geography in the Making*, 323.

54. William O. Field, "Memorandum on the Study of the Variations of Alaskan Glaciers" (n.d.), folder 14, box 18, sub-series b, series 1, Field Papers.

55. François Matthes, "How and When to Measure Advance or Recession," folder 2, box 76, series 7, Field Papers.

56. Gilbert, *Glaciers and Glaciation*, 30.

57. Kate Palmer Albers and Jordan Bear articulate this point using philosopher Charles Saunders Peirce's theory of indexes. An index, according to Peirce, "marks the junction between two portions of experience." Repeat photographs, they assert, mark that junction with special insights. Albers and Bear, "Time Zones," 5.

58. W. Field, "The Glaciers of the Northern Part of Prince William Sound," 376.

59. Odell, "Recent Glaciological Work," 272.

60. Allen, *Life Science in the Twentieth Century*, xv; and Benson, Rainger, and Maienschein, *The American Development of Biology*.

61. Sörlin, "The Anxieties of a Science Diplomat," 70.

62. De Geer, "A Geochronology of the Last 12,000 Years."

63. Sörlin, "The Anxieties of a Science Diplomat."

64. Ryan and Naylor, *New Spaces of Exploration*, 11. See Martin-Nielsen, *Eismitte in the Scientific Imagination*. On Alfred Wegener, see Greene, *Alfred Wegener* and Lüdecke, "Lifting the Veil."

65. From Ahlmann, "The Value and Justification of Polar Research," quoted in Sörlin, "Narratives and Counter-Narratives," 242. See also Ahlmann, "Scientific Results of the Swedish-Norwegian Arctic Expedition," 189.

66. "Glaciological" is specified here because today there are other ways of obtaining mass balance measurements, including using remote-sensing techniques and the hydrological method.

67. See Ahlmann, "Glaciers in Jotunheim and Their Physiography," *Geografiska Annaler* 4 (1922): 50.

68. Ahlmann, "Foreword," 3.

69. Following Gordon Manley, Sörlin describes Ahlmann's glaciers thusly. His argument situates Ahlmann's efforts to build authoritative knowledge within the broader context of site-specific knowledge production that included local knowledge

and assistance plus the establishment of a "school" of glaciological research at Tarfala in the 1940s. The basic elements of Ahlmannian glaciology were, however, already present in others' earlier work and can be understood as part of long-standing efforts to ground the authority of glaciological knowledge in the instrumentalized study of specific places. See Sörlin, "A Microgeography of Authority," 257.

70. Ahlmann, *Land of Fire and Ice*.

71. Jones-Imhotep, *The Unreliable Nation* and Janet Martin-Nielsen, "The Other Cold War."

72. Ahlmann, "Researches on Snow and Ice," 14–23.

73. Ahlmann, "The Present Climatic Fluctuation" and Manley, "Some Recent Contributions," 212. Alas, his theory had negative downstream effects for the research program at Tarfala, where its legacy encouraged skepticism about the possibility that humans could modify the world's climate. See Sörlin, "A Microgeography of Authority," 257.

74. Scientists built alpine research stations on the névés of Swiss glaciers, where they applied instruments and developed techniques for revealing the microstructures of glacier ice to determine exactly how it flowed. The two most important sites were the glacierized Weissfluhjoch and Jungfraujoch. Work at the former was colored by the interests and agenda of mineralogist Henri Bader, who later became a link between Swiss and American glaciology after he moved to the United States. See Bader et al., *Snow and Its Metamorphism*, ix. The Jungfraujoch station was established by British skier and self-taught glaciologist Gerald Seligman, with scientists from Cambridge's Cavendish Laboratory in 1937. Their research emphasized firnification, or the transformation of snow and firn into glacier ice. See Matthes, "Max Harrison Demorest," 13. Glaciology was also progressing in Slavic-speaking lands. Polish glaciologist (or cryologist, to use his term) Antoni Bolesław Dobrowolski wrote the hefty *Natural History of Ice*, advocating for the study of frozen water in all its instantiations under a single science of "cryology." The book was not published until 1923, when it was released in Polish with a French summary. In the Soviet Union Petr Shumskii was studying ice as though it were a rock formation, urging glaciologists to prioritize the processes of glaciation over the phenomena of glaciers; his *Principles of Structural Glaciology* was translated into English in 1964. Dobrowolski's and Shumskii's works were not widely read by North American glaciologists until the 1960s, and by that time bringing Soviet glaciologists to visit the United States proved frustrating and futile. William O. Field to Petr Shumskiy [*sic*], July 30, 1956; William O. Field to Hugh Odishaw, July 2, 1963; and C. Hart to William O. Field, July 7, 1964; all in folder 9, box 15, series 8, Field Papers. The influence of the Swiss research stations was more significant, particularly with Bader's move to America. See D. Inkpen, "Frozen Icons," 127–33; Jürg Sweizer, "Beginning of Snow and Avalanche Research and SLF," accessed January 18, 2018, https://www.slf.ch/en/about -the-slf/portrait/history/beginnings-of-snow-and-avalanche-research-and-the-slf.html; and Haefeli, "The Development of Snow and Glacier Research in Switzerland," 193.

75. Ahlmann, "Present Climatic Fluctuation," 190.

76. See Ahlmann, "The Contributions of Polar Expeditions," 330; and Sörlin, "The Anxieties of a Science Diplomat," 81.

77. W. Field, "Glaciers," 90. Glaciologist Chester Langway referred to Ahlmann and Sverdrup's work in the 1930s as the "beginning of systematic glacier measurements." See Langway, *History of Early Polar Ice Cores*, 5. Patrick Baird and Robert P. Sharp credited Ahlmann with the idea that problems of glacier-climate interactions required "a deeper study of the processes of existing glaciers." See Baird and Sharp, "Glaciology," 29.

3. Measuring

1. Walter Wood to Robert Sharp, October 7, 1947, box 343, folder 1, AGS Records.

2. Doel, "Constituting the Postwar Earth Sciences"; Dennis, "Earthly Matters."

3. Indeed, "what relationships matter within a particular context" is precisely what scales are, according to science studies scholar Max Liboiron. Liboiron, *Pollution Is Colonialism*, 84.

4. François Matthes, "Glacier Measurements in the United States," note on Commission on Glaciers (draft, n.d.), folder 2, box 76, series 7, Field Papers. This idea was echoed by Harry Fielding Reid in "Glaciers and Geophysics," 30.

5. See Goldthwait, "Seismic Sounding."

6. Debarbieux and Rudaz, *The Mountain*; Ferrari, Pasqual, and Bagnato, *A Moving Border*.

7. Sfraga, *Bradford Washburn*, 100; Isserman, *Continental Divide*, 262–66; Putnam, *Green Cognac*.

8. Sfraga, *Bradford Washburn*, 100–145.

9. Field and Brown, *With a Camera in My Hands*, 170–71. The film is available on YouTube: *Land and Live in the Arctic for World War 2 Airmen—1944* (Restored), accessed May 18, 2018, https://www.youtube.com/watch?v=cYwkpDwv5io.

10. "Alaskan Honeymoon: Washburns Are First to Climb Mount Bertha," *Life*, January 20, 1941, 45–54; Miller, "Mount Bertha."

11. When asked to join the 1963 American Mount Everest Expedition, Miller drafted a far-reaching but ultimately untenable plan for geological and glaciological work to add to the team's summit objective. See Clements, *Science in an Extreme Environment*.

12. Miller's was one of three glacier-related projects funded that year, which together accounted for three-quarters of all recipient projects. Fisher, Heald and Miller, "The American Alpine Club Research Fund, 1946," 342.

13. Miller and Tufts geologist Robert Nichols made a reconnaissance mission to Patagonia in early February 1949. They established contacts with Chilean and Peruvian academics and conducted preliminary surveys of some of the glaciers near Lago Argentino and Lae Nahuel Hupi.

14. See William Field to Hans Ahlmann, July 22, 1947, series 8, subseries a, box 1,

folder 1, Field Papers; "Early Draft of Glacier Study Program," n.d., folder 37, box 164, AGS Records; "Statement Outlining a Proposed Glacier Study Project," November 1948, folder 4, box 165, AGS Records; "Memorandum from Dr. Ahlmann," July 15, 1947, series 8, subseries a, box 1, folder 1, Field Papers; Valter Schytt to William Field, February 9, 1954, and Field to Schytt, February 26, 1954, both in folder 43, box 14, subseries a, series 8, Field Papers.

15. See Wolfe, *Competing with the Soviets*; Oreskes, *Science on a Mission*, 17–57; and Oreskes and Rainger, "Science and Security."

16. MacDonald, "Challenges and Accomplishments," 441; and J. Tuzo Wilson, "Arctic Institute of North America," folder 9, box 341, AGS Records.

17. Walter Wood to Ross, September 30, 1948, folder 9, box 341, AGS Records.

18. Wood, "The Parachuting of Expedition Supplies."

19. Robert Sharp to Wood, November 8, 1948, folder 1, box 343, AGS Records.

20. Robertson, "The First Ascent of Mount Vancouver."

21. Sharp confessed to not knowing why Wood asked him to join the 1941 expedition or how he learned of his name. Sharp to LaChapelle, July 21, 1953, folder 1, box 9, Sharp Papers.

22. Odishaw to Sharp, June 13, 1957, folder 33, box 2, Sharp Papers.

23. Sharp to Field, May 8, 1959, folder 8, box 5, Sharp Papers.

24. Campbell to Edward LaChapelle, July 15, 1953, folder 1, box 9, Sharp Papers; Robert Sharp, "What Is Caltech?," October 24, 1959, folder 25, box 36, Sharp Papers.

25. Field to Sharp, May 22, 1951, folder 1, box 15, subseries a, series 8, Field Papers.

26. David A. Valone, interivew of Clarence R. Allen, Archives of the California Institute of Technology, April 1 and 4, 1994, http://archives.caltech.edu/repositories/2 /accessions/165.htmlfrom=o&q=Allen%2C+Clarence&q _exact=&q_excluded=&q _required=&size=10&totaı=69.

27. Sharp to Field, November 17, 1947, and Sharp to Field, June 15, 1948, folder 2, box 15, sub-series a, series 8, Field Papers.

28. Sharp to Wood, October 16, 1947, folder 1, box 343, AGS Records.

29. Ice flow was a hot topic. Before his premature death on the Greenland ice cap in 1941, American glaciologist Max Demorest was generating excitement for being the first theoretical glaciologist possibly equal to European caliber. This was largely due to his account of extrusion flow, which was simultaneously theorized by Rudolf Streiff-Becker. Extrusion flow was a form of glacier movement that scientists thought would occur when, given sufficient mass, pressure from upper ice layers would cause lower ones to flow outward from underneath (imagine the contents of a sandwich oozing out the sides when the top piece of bread is pressed down). Extrusion would likely occur when the ice was not bounded at its margins, as on large ice caps such as Greenland's. Sharp wanted to know if there was evidence of extrusion flow on piedmont glaciers like Malaspina, which also flow unhindered across relatively flat topographies. His measurements showed that the upper layers moved faster than the lower ones, providing evidence

against extrusion flow. During the same years, British theorist John Nye dealt the extrusion flow theory a death blow using a mathematical argument that showed the forces required were impossible to produce. See Waddington, "The Life, Death and Afterlife," 989; Nye, "The Flow of Glaciers"; and Nye, "The Mechanics of Glacier Flow."

30. Sharp to Wood, October 16, 1947, and Sharp to Wood, October 22, 1947, folder 1, box 343, AGS Records.

31. Robert P. Sharp, "Glaciological Studies on the Seward Ice Field, St. Elias Range, Alaska-Canada Status Report," November 1, 1948, folder 9, box 341, AGS Records; and Sharp to Wood, March 30, 1950, folder 1, box 343, AGS Records.

32. Maynard Miller to Field, June 19, 1948, folder 1, box 164, AGS Records; Maynard M. Miller, "The Juneau Ice Field Research Project, Report of 1948 Field Work," December 1948, folder 44, box 164, Field Papers.

33. His 1948 glaciological survey of the icefield also included a climber's assessment of Devils Paw (2,616 m). The four-pronged fist of granite sparked his imagination, instigating plans to return to it the following summer. His ambitions to stand atop its summit led to trouble the following year between himself and famed mountaineer Fred Beckey. D. K. Inkpen, "Scientific Life in the Alpine."

34. The Seward Icefield, from which the Malaspina flows, straddles the Alaska-Yukon border, which meant there was a question of which nation could and would provide air support. The closest air base was the American one in Yakutat. Ultimately, proximity decided the matter. Wood to Sharp, October 7, 1947, folder 1, box 343; "Project Fairy Castle," n.d., folder 15, box 341; Wood to Jackman, October 7, 1947; and Wood to Lemmons, October 7, 1947, folder 8, box 345, all in AGS Records.

35. Sapolsky, *Science and the Navy*, 37.

36. Sfraga, *Bradford Washburn*, 160.

37. Howkins, "Science, Environment, and Sovereignty"; Doel et al., "Strategic Arctic Science"; and Doel, Harper, and Heymann, *Exploring Greenland*.

38. Jones-Imhotep, *The Unreliable Nation*, 35. See also Piper, "Introduction"; Doel, Wråkberg, and Zeller, "Science, Environment, and the New Arctic"; Lackenbauer and Farish, "The Cold War on Canadian Soil"; and Korsmo, "The Early Cold War."

39. Dennis, "Earthly Matters," 809.

40. The literature exploring science and geopolitics in the polar regions is substantial. Notable studies include Dodds, *Geopolitics in Antarctica*; O'Reilly, *Technocratic Antarctic*; Powell, "Lines of Possession?"; Powell, "Science, Sovereignty and Nation"; and Powell and Dodds, *Polar Geopolitics?*

41. Martin-Nielsen, "The Other Cold War."

42. In addition to the works cited in previous notes, Adrian Howkins's work on the IGY is particularly instructive in its attention to geopolitical details. See Howkins, "The Significance of the Frontier"; and Howkins, "Reluctant Collaborators." The geopolitical stakes were also made apparent in China's non-involvement. See Wang and Zhang, "China and the International Geophysical Year."

43. Field, "Statement Outlining a Proposed Glacier Study Project," 52; Robert P. Sharp,"Outline of Glaciological Program for St. Elias Range Research Station," n.d., folder, 15, box 431, AGS Records.

44. See Chu, "Mapping Permafrost Country."

45. Heggie, *Higher and Colder*, 66.

46. Wood, "The Parachuting of Expedition Supplies," 38–40.

47. Wood to Andrew McNaughton, February 11, 1948, and Charles Deerwester to Wood, March 12, 1948, folder 8, box 345, AGS Records.

48. Maynard M. Miller, "The Juneau Ice Field Research Project, Report of 1948 Field Work," folder 44, box 164, Field Papers.

49. Calvin Heusser to Field, July 6, 1952, folder 31, box 166, AGS Records.

50. Oreskes and Conway, *Merchants of Doubt*; Brandt, *The Cigarette Century*; Proctor, *Golden Holocaust*.

51. Naomi Oreskes has shown this to be the case also for Cold War physical oceanographers and marine geophysicists. See Oreskes, *Science on a Mission*.

52. Edwards, *The Closed World*.

53. Oreskes, *Science on a Mission*.

54. Hamblin, *Arming Mother Nature*.

55. Oreskes, *Science on a Mission*, 480.

56. Miller, "Twenty-Five Explorers." The plane belonged to Snow Cornice, which Miller had visited that summer, but Miller's irresponsibility with classified material upset Field nonetheless. Field to Sharp, March 9, 1949, folder 2, box 15, subseries c, series 8, Field Papers; Field to Miller, February 28, 1952, and Miller to Field, March 3, 1952, all in box 19, subseries a, series 8, Field Papers. Miller's greatest slight against his superiors and sponsors was his later refusal to provide them with a copy of his doctoral thesis, which contained data from his seasons on the Juneau, folder 14, box 10, subseries a, series 8, Field Papers.

57. The *American Alpine Journal* and the *Canadian Alpine Journal* continued to be outlets for publishing glacier research into the 1960s. See Fisher, "Studies of Forbes Dirt Bands"; Mathews, "Glacier Study for the Mountaineer"; and Patterson, "Glaciological Research in Western Canada."

58. The extent to which the military demanded such information changed over the years. Early on, the ONR demanded reports on logistics and equipment from Walter Wood; by the mid-1950s the ONR was calling for data on more of the technical work on snow and ice done by men like Edward LaChapelle and Richard Hubley. Still, in 1956, LaChapelle provided write-ups on avalanches and glacier travel for a report a glaciers and glaciation contracted by the Quartermaster Corps. LaChapelle, "Some Aspects of Glacier Travel with Particular Reference to Military Operations," February 18, 1957, folder 2, box 8, subseries a, series 8, Field Papers.

59. Miller to Field, August 10, 1948, folder 40, box 164, AGS Records; early draft of "Glacier Study Program," n.d., 3, folder 37, box 164, AGS Records.

60. "Statement Outlining a Proposed Glacier Study Project," November 1948, folder 4, box 165, AGS Records.

61. "Juneau Icefield Research Project" film reels (1948–58), University of Wisconsin–Milwaukee Libraries, accessed October 25, 2017, http://collections.lib.uwm.edu /digital/collection/agsnorth/id/13566.

62. Jirps used aerial photographs to map the icefield. In the project's final years, after relocating to the Lemon Creek Glacier, they used them to document changes in the firn line, which gave an approximation of the glacier's regimen. More about that is discussed in the next chapter.

63. Roberts, "Walter Wood and the Legacies."

64. Field to Sharp, February 27, 1953, folder 5a, box 5, Sharp Papers.

65. JIRP Semi-Annual Status Report, December 1956; JIRP Semi-Annual Status Report, June 1956; JIRP Semi-Annual Status Report, November 1957; JIRP Semi-Annual Status Report, November 26, 1958, folder 19, box 155, AGS Records.

66. Task Order N9onr 83001, amendments 1–10, folder 18, box 155, AGS Records.

67. The Juneau Icefield Research Project wrapped up in 1958. The Juneau Icefield Research Program, with a focus on education, was begun by Mal Miller shortly thereafter. While Miller claimed continuity between the two endeavors, his colleagues at the AGS would have disagreed. In terms of funding, institutional oversight, personnel, and aims, they were very different. The Juneau Icefield Research Program continues to this day and continuity with the original project is claimed in institutional histories. See Andrew Opila, "What Is JIRP?" accessed September 23, 2021, https://juneauicefield .org/; and "Obituary: Maynard Malcom Miller," at Juneau Icefield Research Program, accessed September 23, 2021, https://juneauicefield.org/blog/tag/Maynard+Miller.

68. Heusser to Charles Hitchcock, June 30, 1958, folder 19, box 155, AGS Records.

69. Daywire DA1456, folder 9, box 345, AGS Records.

70. Sharp to Christopher Ehrendreich, June 19, 1952, folder 45, box 15, Sharp Papers.

71. Sharp to Wood, January 3, 1952; Sharp to Wood, May 27, 1952; and Sharp to Wood, April 9, 1953, all in folder 1, box 343, AGS Records.

4. Monitoring

1. Walter Sullivan, "Dr. Hubley Dies at Glacier Camp," *New York Times*, October 31, 1957, 28; "Hubley's Arctic Death Ascribed to Suicide," *Washington Post and Times Herald*, November 3, 1957, B12.

2. Wood, "Richard Carleton Hubley."

3. In a letter written shortly after Hubley's death, Edward LaChapelle confessed to Sharp that he knew from Austin Post that "Dick" had been struggling with "some rather serious personal difficulties" and had sought psychiatric help. Edward LaChapelle to Robert Sharp, November 1957, box 2, folder 33, Sharp Papers.

4. Fleagle, *Eyewitness*, 14–17.

5. Post also left his own IGY project to help complete the McCall work. Yet Post was no micrometeorologist and Hubley's results were left unanalyzed for forty-five years. Weller et al., "Fifty Years of McCall Glacier Research," 103.

6. Warde, Robin, and Sörlin, *The Environment*, 24. Etienne Benson employs the term "surroundings" to capture the ways that ideas of environment have expanded and contracted to include interior environments, such as that of our gut or urban geographies, and to connect the idea of environment to earlier conceptualizations and concerns about the worlds in which we and other organisms abide. Benson, *Surroundings*.

7. Warde, Robin, and Sörlin, *The Environment*; McNeill, *Something New Under the Sun*; Warde, "The Environment."

8. See Kuklick and Kohler, *Science in the Field*; Kohler, *Landscapes and Labscapes*; Vetter, *Knowing Global Environments*; Kohler and Vetter, "The Field."

9. I borrow the term "generalizable" from science studies scholar and environmental scientist Max Liboiron. They, following Eve Tuck and Wayne Yang, have argued that scientists would do better to seek generalizable "relational validity" rather than simply assume that scientific knowledge is universally valid. Liboiron, *Pollution Is Colonialism*, 154. On the difficulties of reproducing lab results (both practical and philosophical) see Collins, "The TEA Set"; Collins, "Reproducibility of Experiments."

10. On science as a place-based activity, see Livingstone, *Putting Science in Its Place*, and Shapin, "Placing the View from Nowhere."

11. LaChapelle to William Field, October 3, 1954, folder 1, box 8, subseries a, series 8, Field Papers.

12. Richard Hubley to Sharp, December 28, 1956, folder 47, box 7, Sharp Papers.

13. Hubley, "Recent Growth of Glaciers," 162.

14. LaChapelle, "Assessing Glacier Mass Budgets," 290.

15. Sullivan, "Dr. Hubley Dies," 28; Pfeffer, "Obituary: Austin Post": 28–33; Ives, "Vanishing Uplands."

16. Mark F. Meier, "Project Proposal: Glaciers of the North Cascades," September 4, 1958, folder 15, box 10, Sharp Papers.

17. Charles Morrison to Post, September 1960, folder 5, box 13, subseries a, series 8, Field Papers.

18. W. Tad Pfeffer, "Austin Post," unpublished manuscript, 28.

19. Post's biographical details are taken from Scurlock, "Austin Post"; Anonymous, "A Lifelong Love Affair with Glaciers," *Seattle Times*, November 22, 2012; Austin Post, "A Summer with the Coast and Geodetic Survey in Alaska, 1949," unpublished, n.d., courtesy of David Janka.

20. Author interview of Wendell Tangborn, Erin Whorton, and Birdie Krimmel, June 23, 2017, Vashon Island, Washington.

21. Heusser, *Juneau Icefield Research Project Retrospective*, 95.

22. Austin Post to Field, January 4, 1980, folder 3, box 13, subseries a, series 8, Field Papers.

23. Pfeffer, "Austin Post," 2.

24. A 1958 IGY briefing in *Science* devoted two full pages to Antarctica and three paragraphs to glaciological programs in the Northern Hemisphere. See also Odishaw, "International Geophysical Year."

25. LaChapelle to Sharp, November 23, 1955, folder 1, box 9, Sharp Papers.

26. Walter A. Wood, "Proposal for the Long-term Study of a Glacierized Geographical Area," December 1960, folder 45, box 15, Sharp Papers.

27. There were four main field studies in the US's Northern Hemisphere glaciological program: Sharp's velocity studies at the Blue Glacier; Edward LaChapelle's regimen studies on the upper Blue Glacier; Richard C. Hubley's micrometeorological and physical studies on the McCall Glacier; and Mark Meier's multifaceted project on the South Cascade Glacier. Contributions from ongoing projects on glaciers of the Northwest, the Rockies, and the Lemon Creek Glacier (JIRP) were also included. Mark F. Meier, "Glaciological Activities During the IGY-IGC: The United States Exclusive of Alaska," folder 1, box 19, subseries a, series 8, Field Papers.

28. Post to Field, March 16, 1959, folder 5, box 13, subseries a, series 8, Field Papers.

29. Post to Brown, n.d.; Austin Post, "Progress Report," September 1, 1960, folder 1, box 13, subseries a, series 8, Field Papers; Post to Field, March 16, 1959, folder 5, box 13, subseries a, series 8, Field Papers.

30. Meier to Field, n.d., box 156, folder 4, AGS Records; quoted in Hitchcock to Post, December 10, 1957, box 156, folder 4, AGS Records.

31. Post, "The Exceptional Advances."

32. Meier to Sharp, October 17, 1958, folder 15, box 10, Sharp Papers.

33. It was Meier who pointed out the Saskatchewan Glacier's potential to Sharp, when the latter was tired of working on the Malaspina. Meier in turn was steered toward the Saskatchewan by Field, having recently visited the Saskatchewan in 1948 and 1949. Meier to Field, October 18, 1951, folder 6, box 26, series 4, Field Papers. Meier's work in 1952 was supported by ONR contract N8onr-367, administered as a grant-in-aid through AINA, and in 1953 by ONR contract N6onr 25516, NR081-069, which was directly administered through Caltech. The project's full name was "A Study of the Lower Course of the Saskatchewan Glacier, Alberta, Canada," Project ONR-89, Task Order 1, Contract N8onr-367. Sharp to Colbert, November 21, 1952, and Sharp to Colbert, June 21, 1954, folder 32, box 1, Sharp Papers; Chief of Naval Research to Chief of Naval Operations, December 24, 1952, folder 16, box 13, Sharp Papers.

34. Sharp to Meier, April 24, 1956, folder 15, box 10, Sharp Papers.

35. Tangborn, Whorton, and Krimmel interview. Author phone interview of Andrew Fountain, October 30, 2017. "Draft Notes on Meier for Fulbright Application," n.d.; "Meier Fulbright application letter"; and Meier to Sharp, June 17, 1957, all B10, F15, Sharp Papers.

36. Goetzmann, *Exploration and Empire*, 573–78, 582, 595.

37. Memo, Raymond Nace to Branch Chiefs, Water Resources Division, September 20, 1956, folder 15, box 10, Sharp Papers.

38. Meier, "Research on the South Cascade Glacier," 41.

39. Tangborn, Krimmel, and Meier, "A Comparison of Glacier Mass Balance; Meier, "The South Cascade"; Fountain and Funk, "South Cascade Glacier Bibliography."

40. Post, "Alaskan Glaciers."

41. Scurlock, "Obituary: Austin Post."

42. The magisterial gaze is an elevated perspective on a landscape, typically a mountain one. Conjuring ideas of manifest destiny and the western frontier, there is almost always a gap or way through that leads the eye over and through the mountains. In glacier photographs the road-like strip of white, the glacier itself, serves the same function of leading the eye through the mountains. Boime, *The Magisterial Gaze.*

43. Bruns, "Grieving Okjökull."

44. W. O. Field, "Comment to NSF on Proposal P-11748 for Austin Post Air Photography of Glaciers," June 1, 1962, folder 1, box 13, subseries a, series 8, Field Papers. See also Schytt, "A. (S.) Post and E. R. LaChapelle."

45. Post and LaChapelle, *Glacier Ice,* xii.

46. Pfeffer, "Austin Post," 2–3.

47. Carter, "Icebergs and Oil Tankers"; NOAA, "Oil Spill Case Histories, 1967–1991," 80.

48. This ratio was calculated by dividing the area above the firn line by the total area of the glacier.

49. LaChapelle, "Assessing Glacier Mass," 297. See also Meier, "Mass Budget of South Cascade Glacier"; and Shumskii, "The Energy of Glaciation."

50. Post and LaChapelle, *Glacier Ice,* 107.

51. During the 1970s the idea of an oncoming ice age circulated in popular media and a few scientific papers. The idea was, however, in the minority among peer-reviewed articles on the topic of global climate, the majority of which, according to a scientists' review of the literature, were concerned with the possibility of warming. See Peterson, Connolley, and Fleck, "The Myth of the 1970s."

52. Howe, "This Is Nature," 286.

53. See Edwards, *Vast Machine.*

54. Weart, *The Discovery of Global Warming,* 57–59, 107–16.

55. Joshua Howe points out that models were more convincing to some scientists than others. Historical geologists, climatologists, and meteorologists preferred "physical, documentary and statistical evidence"; they were not as easily convinced by the models. See Howe, *Behind the Curve,* 93–100.

56. "Carbon Dioxide and Climate," iv.

57. Edwards, "Meteorology as Infrastructural Globalism"; Edwards, *Vast Machine,* 8–25.

58. Boudia, "Observing the Environmental Turn," 196.

59. Vetter, *Knowing Global Environments*; Hamblin, *Arming Mother Nature.*

60. To get a sense of the focus on oceanography and meteorology, see Weart, *The Discovery of Global Warming,* 57–59, 107; Fleming, *Historical Perspectives*; Howe, *Be-*

hind the Curve, 103–17; Oreskes and Conway, *Merchants of Doubt*, 169–85; and Howe, *Making Climate Change History*.

61. Ohmura, "Completing the World Glacier Inventory."

62. Boudia, "Observing the Environmental Turn"; Edwards, "Meteorology as Infrastructural Globalism."

63. Taylor et al., "Remote Sensing," 2; Wallén, "Monitoring the World's Glaciers"; Luthcke et al., "Antarctica, Greenland."

64. Longino, "Evidence and Hypothesis," 44.

65. Glaciologists had developed ways of estimating volume from surface images, but these were not very accurate. According to Ethan Welty, a photographer and glaciologist at the University of Colorado, even digital photography, which includes metadata (i.e., orientation, location by GPS, timing, etc.) still requires "photogrammetic calisthenics" to gain some measure of control. See Conway, Rasmussen, and Marshall, "Annual Mass Balance of Blue Glacier."

66. See Sörlin, "The Anxieties of a Science Diplomat," 66–88; Sörlin, "The Global Warming," 93–114.

67. Tangborn, Whorton, and Krimmel interview. "I was watching the glacier balance every year," he told me, "and it seemed like all of a sudden they were always negative. And before that they'd go back and forth. And 1980 was a year that I remember when they were just predominantly negative. The first year, first time. And it just seemed so unusual."

68. Þórarinsson, "Present Glacier Shrinkage."

69. Mercer, "West Antarctic Ice Sheet."

70. This brief account of WAIS research is indebted to the excellent work of Jessica O'Reilly and her coauthors. See Oppenheimer et al., "Assessing the Ice," 132; and O'Reilly, Oreskes, and Oppenheimer, "A Rapid Disintegration of Projections."

71. Half of that increase occurred between 2003 and 2009. See Mouginot, Rignot, and Scheuchl, "Sustained Increase in Ice Discharge."

72. See Weart, *The Discovery of Global Warming*, 79–80; Hughes, "John H. Mercer (1922–1987)"; Thomas, "Research Agendas in Climate Studies"; and Oppenheimer et al., "Assessing the Ice," 132.

73. Robin, "Review of *Sea Level, Ice, and Climatic Change*."

74. Young, "Responses of Ice Sheets," 355. See also Flohn, "Climatic Change, Ice Sheets and Sea Level." The Maine conference was organized by University of Maine glaciologists Terence Hughes and Jim Fastook and was sponsored by the American Association for the Advancement of Science and the US Department of Energy. Oceanographer Roger Revelle, though not present, was a key instigator.

75. "Provisional Agenda, Workshop on Glaciology," June 11, 1980, folder 16, box 24, subseries b, series 8, Field Papers. In 1981, the report "Environmental Consequences of CO_2-induced Climatic Change" was part of a meeting of the US Committee on Glaciers. Committee member Charley Bentley's paper, "CO_2 Warming and West Antarc-

tica," was included in the "New Fields, Change of Emphasis" section *of the 1981 report.*

76. Oppenheimer et al., "Assessing the Ice."

77. Oppenheimer et al., *Discerning Experts.* For a different approach to assessment work, one that tries to cultivate good relations with Indigenous knowledge-holders, see the Canadian Mountain Assessment at https://www.canadianmountainnetwork.ca /research/canadian-mountain-assessment-group/canadian-mountain-assessment.

78. Phone interview of W. Tad Pfeffer, February 20, 2018.

79. Recent work suggests another mechanism by which West Antarctica might contribute to sea level rise. Harvard University glaciologists found in 2021 that bedrock uplift—occurring because the ice sheets are thinning—would push remaining ice and water into the oceans. This is called the water-explosion mechanism, and it could add an additional meter of water to the seas over the next thousand years were the ice sheets to collapse. See Pan et al., "Rapid Postglacial Rebound."

80. Jessica O'Reilly refers to such data as "charismatic." See O'Reilly, *The Technocratic Antarctic*, 160–70.

81. The question was posed to him by Scripps oceanographer Tim Barnett. See Barnett, "Recent Changes in Sea Level."

82. Pfeffer, "Land Ice and Sea Level Rise," 98.

83. Meier, "Snow and Ice," 5.

84. This take on consensus comes from Oppenheimer et al., *Discerning Experts*, 10–12.

85. National Research Council, *Glaciers, Ice Sheets, and Sea Level*, 1–6.

86. Working with an imperfect and incomplete data set, Meier estimated that for the period of 1900–1961, the retreat of small glaciers contributed 2.8 cm, or 0.46 (+/-0.26) mm, per year to global sea level rise. He had litle data on the magnitude of long-term changes. To make his calculation, he had to assume that long-term changes were proportional to the difference between winter and summer mass balances, which allowed him to extrapolate from measurements on a few well-documented glaciers. Using data from the South Cascade and Sarennes Glaciers and Storglaciaren he extrapolated the record to include the period between 1885 and 1974, which gave an average rate of sea level rise per year. See Meier, "Contributions of Small Glaciers."

87. Field, scientific review for *Science*, May 22, 1984, folder 1, box 19, subseries a, series 8, Field Papers.

88. Warrick and Oerlemans, "Sea Level Rise," 268–69. To see how IPCC assessment reports dealt with sea level projections and the WAIS, see Oppenheimer et al., *Discerning Experts*, 127–69; O'Reilly et al., "A Rapid Disintegration."

89. Meier, "Role of Land Ice"; Meier, "Ice, Climate, and Sea Level"; Meier and Wahr, "Sea Level Is Rising"; Meier et al., "Glaciers Dominate."

5. Witnessing

1. John F. Shroder Jr., "2009 GSA Public Service Award: Presented to Bruce F. Molnia," accessed July 25, 2021, https://www.geosociety.org/awards/09speeches/psa.htm.

2. Anecdotally, another glaciologist was asked about glacier photographs by a government official in Washington state around this time. Author conversation with Michele Koppes, May 26, 2022, Banff, Alberta.

3. Terrell Johnson, "Bruce Molnia's Repeat Photos of Alaska, and What He Says They Reveal About Our World," The Weather Channel, July 18, 2014, https://weather.com/science/environment/news/alaskas-glaciers-capturing-earth-changing-our-eyes-20131125.

4. Explorers Club Washington Group, "Bruce Molnia," accessed June 23, 2022, https://www.explorersclubdc.org/about-2/officers-and-board-members/bruce-molnia/.

5. For a small sampling of this discourse see, Appenzeller, "The Big Thaw"; USGS, "Most Alaskan Glaciers Retreating, Thinning, or Stagnating," *Science News*, October 6, 2008; USGS, "Repeat Photography Project," accessed September 13, 2015, https://www.usgs.gov/centers/norock/science/repeat-photography-project?qt-science_center_objects=0#qt-science_center_objects; Mauri Pelto, "From a Glacier's Perspective," accessed September 13, 2015, https://glacierchange.wordpress.com; Hassan Basagic, "Basagix Field Notes," accessed September 13, 2015, http://hassanbasagic.blogspot.com/2007/11/glacier-measurements.html; Balog, *Ice*; Balog, *Extreme Ice Now*; Johnson, "Bruce Molnia's Repeat Photos"; Mark Hume, "Disintegrating Rockies Glacier Sends 'Strong Message' on Climate," *Globe and Mail*, October 22, 2016, https://www.theglobeandmail.com/news/british-columbia/disintegrating-rockies-glacier-sends-strong-message-on-climate/article26945443/; Keith Bradshear, "The Paris Agreement on Climate Change Is Official, Now What?," *New York Times*, November 3, 2016, https://www.nytimes.com/2016/11/04/business/energy-environment/paris-climate-change-agreement-official-now-what.html; Associated Press, "Then and Now: Photographers Document Rapid Melting of World's Glaciers," CBC, April 3, 2017, http://www.cbc.ca/news/technology/then-and-now-world-glaciers-melting-1.4053208; Justin Gillis, "Earth Sets a Temperature Record for the Third Straight Year," *New York Times*, January 18, 2017, https://www.nytimes.com/2017/01/18/science/earth-highest-temperature-record.html; Nadja Popovich, "Mapping 50 Years of Melting Ice in Glacier National Park," *New York Times*, May 24, 2017, https://www.nytimes.com/interactive/2017/05/24/climate/mapping-50-years-of-ice-loss-in-glacier-national-park.html; Kintisch, "Meltdown."

6. This sentiment was voiced by veteran University of Alaska glaciologist Carl Benson, in conversation with the author on July 24, 2014. Bill Field certainly voiced feelings of alienation in line with Benson's assessment. In 1979 he complained to Post that glaciology wasn't "the same as that with colleagues that tramp the mountains with camera

in hand." Austin Post to William Field, December 3, 1979, folder 4, box 13, subseries a, series 8, Field Papers.

7. Molnia interview.

8. A helpful primer for getting a better sense (from just the title, even) of the historian's science is Steven Shapin's *Never Pure: Historical Studies of Science as if It Was Produced by People with Bodies, Situated in Time, Space, Culture, and Society, and Struggling for Credibility and Authority.*

9. "Public" is here just a shorthand term for nonscientist, and even more specifically, non-climate-related-scientist. It's a loose way of drawing a boundary around a group of experts. Of course, there is no single, unified group that is *the* public. Publics are always shifting and exist only relative to the group of experts in question. Indeed, scientists are part of a public when untutored in an area of relevance. For example, a quantum astronomer is part of a nonscientific public when it comes to ecological knowledge because she is not an expert in that area, and vice versa for the ecologist who finds herself in a quantum observatory. See Warner, *Publics and Counterpublics.*

10. See Oreskes, "Of Mavericks and Mules."

11. IPCC, *Climate Change 1995*, 5. The IPCC's 2001 Synthesis Report went even further, stating "New and stronger evidence that most of the warming observed over the last 50 years is attributable to human activities." Watson and Core Writing Team, *Climate Change 2001*, 5. See also Oreskes, "Beyond the Ivory Tower."

12. Callison, *How Climate Change Comes to Matter*, 3; "Americans' Global Warming Concerns Continue to Drop," *Gallup News*, March 11, 2010, https://news.gallup.com /poll/126560/americans-global-warming-concerns-continue-drop.aspx. See also Oreskes and Conway, *Merchants of Doubt*, 169.

13. Carey, "The Trouble with Climate Change"; Callison, *How Climate Change Comes to Matter*; Andrew Stuhl, *Unfreezing the Arctic*; Carey, *In the Shadow of Melting Glaciers*; Howe, *Behind the Curve.*

14. Jackson, *The Secret Lives of Glaciers*; Gagné, *Caring for Glaciers.*

15. Hulme, *Why We Disagree About Climate Change*, xxvi.

16. Scholars disagree over whether climate change is invisible or visible with unaided senses. Peter Rudiak-Gould points out that this question is partly an empirical question of how dramatically humans have tampered with Earth systems, partly an instrumental question of how to communicate anthropogenic climate change, and partly "a moral, political, and epistemological interrogation of the value and role of scientific expertise in democratic society" (129). To ask if climate change is visible or invisible is to ask whom we ought to trust. Rudiak-Gould, "'We Have Seen It."

17. Early on, many advocates assumed things would proceed as they had for ozone-degrading chlorofluorocarbons (CFCs), the chemicals widely used in aerosols, packing materials, solvents, and refrigeration. In 1974 chemists showed that CFCs could decompose into ozone-destroying chlorine when exposed to radiation. Ozone, a trace gas found predominantly in the stratosphere, protects the living world below from

harmful UV rays. Humans need ozone to avoid very, very bad sunburns, and worse. In 1985 a large "hole" in the ozone layer was identified hovering over Antarctica, precipitating widespread calls to regulate CFCs. Industry pushed backed against regulation, but any denial of CFCs' environmental impacts was short-lived. Two years later the Montreal Protocol, an agreement to phase out ozone-depleting emissions to pre-1980s levels, was signed by twenty-four nations with relatively little political opposition. It seemed a tidy case of identifying a problem, ascertaining its causes, and agreeing to regulate those causes. The United Nations Framework Convention on Climate Change was modeled on this experience, but global warming proved to be a more complicated, trenchant issue to address. Weart, *The Discovery of Global Warming*, 152–54.

18. "Smoking and Health Proposal," 1969, BN: 680561778, Legacy Tobacco Documents Library, quoted in Oreskes and Conway, *Merchants of Doubt*, 34.

19. An alternative strategy, Howe suggests, would have been to shift the focus from the certainty of the science to the covert political and ideological ideas that informed global warming debates: to confront the value-saturated nature of the policy discussions directly, exposing merchandizing of doubt for what it was, and recognizing the values informing scientists' positions. Howe, *Behind the Curve*, 9, 194.

20. Boykoff, *Who Speaks for the Climate?*

21. Merton, *The Sociology of Science*; Brysse et al., "Climate Change Prediction."

22. Brysse et al., "Climate Change Prediction," 333–34.

23. Daston and Galison, *Objectivity*, 35.

24. Feminist science studies scholars have delved deeply into the ways drama does not sit well with stereotypical notions of scientific objectivity and the effect this has had on the gendering of science. See Laslett et al., *Gender and Scientific*; Haraway, *Primate Visions*; Rossiter, *Women Scientists in America*; and Harding, *The Science Question in Feminism*.

25. Molnia interview.

26. Brysse et al., "Climate Change Prediction," 334; Oppenheimer et al., *Discerning Experts*, 127-69; Schneider, *Science as a Contact Sport*.

27. Gieryn, "Boundary-Work"; Rudwick, "The Emergence of Visual Language," 177, 179–81.

28. Jasanoff, "A New Climate for Society," 235; Finis Dunaway, "Seeing Global Warming," 10.

29. Orlowski, *Chasing Ice*, timestamp 10:03.

30. This account of the limits of representation leans heavily on the work of philosopher Michael Weisberg, according to whom ideals that guide the construction of representations cannot, for logical and practical reasons, be maximally realized in any single representation. Weisberg, *Simulation and Similarity*, 98–112. Weisberg is primarily interested in models; however, his account can be applied to other kinds of scientific representations, including visual representations and experiments. In the latter case see, for example, S. A. Inkpen, "Like Hercules and the Hydra."

31. See Borges, *On Exactitude in Science*: "In that Empire, the Art of Cartography attained such a Perfection that [...] the Cartographers Guild struck a Map of the Empire whose size was that of the Empire, and which coincided point for point with it. The following Generations, who were not so fond of the Study of Cartography as their Forbears had been, saw that the vast Map was Useless" (325).

32. Molnia interview.

33. Pfeffer, *The Opening of a New Landscape*, xi.

34. Pfeffer, "Glaciology Needs to Come Out."

35. Pfeffer, *The Opening of a New Landscape*.

36. Author phone interview with W. Tad Pfeffer, February 26, 2018.

37. This pattern follows Joseph Campbell's classic articulation of the hero's journey to a tee. See Campbell, *The Hero with a Thousand Faces*.

38. Jackson, "Glaciers and Climate Change."

39. In Canada, the Mountain Legacy Project, backed by a consortium that includes academic, government, and private supporters and led by ecologist Eric Higgs (University of Victoria), sends teams of student researchers into the Canadian Rockies every summer to recreate the historic photographs taken by the Vauxes, Mary Schäffer, Bill Field, and their contemporaries. Many of their images are of glaciers, though these are not the primary focus of the program.

40. Associated Press, "Then and Now: Photographs Document Rapid Melting of World's Glaciers," *CBC News*, April 3, 2017, https://www.cbc.ca/news/science/then-and-now-world-glaciers-melting-1.4053208.

41. This is called photographic realism. See Wilder, *Photography and Science*, 102–9; Daston and Galison, *Objectivity*; J. Snyder, "Res Ipsa Loquitur," 221. Scholars of visual culture have warned against a facile trust in the transparency of photographs, yet the "truth" of photography remains compelling as a common-sense assumption even, ironically, at a time when digital manipulation makes it less straightforwardly obvious. Thankfully, it is not our task here to judge belief in photographic realism. Deciding whether or not advocates of repeat photography were right to assume that "seeing is believing" is above and beyond the task of understanding why they did. Flusser, *Toward a Philosophy of Photography*, 15; Sontag, *On Photography*.

42. Tufte, *Visual Explanations*, 73–77.

43. The classic articulation of this thesis is Nicholson, *Mountain Gloom and Mountain Glory*. See also Carey, "The History of Ice."

44. Williams, "Ideas of Nature"; Daston and Vidal, *The Moral Authority of Nature*; Cronon, "The Trouble with Wilderness."

45. The Habitable Planet, interview of Lonnie Thompson, "The Earth's Changing Climate," unit 12, Annenberg Foundation, https://www.learner.org/courses/envsci/unit/text.php?unit=12&secNum=0.

46. Rumore, "A Natural Laboratory."

47. Here I echo the introduction to Lorraine Daston and Peter Galison's mighty history of scientific objectivity, *Objectivity*, 17.

48. Shapin, *Never Pure*.

49. Similarly, Naomi Oreskes has shown that for geologists, predictions "serve a role that is primarily social, rather epistemic." Geologists' turn to predictive modeling in the 1960s and 1970s was a divergence from a long history in which explaining the past was more important than predicting the future. This shift, she contends, was done in service of Cold War military patrons. Oreskes, "From Scaling to Simulation," 119–20.

Conclusion

1. I am not alone in this assessment. Writer Haruki Murakami, who is also a long-distance runner and has trained in many places, including Hawaiʻi, finds Boston heat especially brutal. Murakami, *What I Talk About*.

2. Thank you, Steven Shapin. I am paraphrasing, and you probably don't remember this. But it helped.

3. W. Tad Pfeffer phone interview, February 26, 2018.

4. Steig, "Another Look," 5–7.

5. Carey, "The Trouble with Climate Change," 259.

6. Marzeion et al., "Attribution of Global Glacier Mass," 919.

7. Cronon, "Trouble with Wilderness"; Cosgrove, "Images and imagination."

8. Jackson, "Narratives"; Dunaway, *Seeing Green*, 2; Carey, "The Trouble with Climate Change," 258.

9. Liboiron, *Pollution Is Colonialism*. Whyte theorizes settler colonialisms as being structured on oppressions based on one society's interference with and erasure of another society. He argues that environmental injustice includes ways that one society "robs" another of its capacities to "experience the world as a place of collective life that its members feel responsible for maintaining into the future." Whyte, "Indigenous Experience," 158. See also Whyte, "Is It Colonial Déjà Vu?"; Whyte, "Indigenous Science (Fiction)"; Bennett et al., "Indigenous Peoples, Lands, and Resources."

10. Hulme, *Why We Disagree*.

11. Sörlin, "Cryo-History," 327.

12. For a small sampling, see Tsing et al., *Arts of Living on a Damaged Planet*; Schmidt and Wolfe, *Climate Change*.

13. Compare Stibal, Šabacká, and Žárský, "Biological Processes."

14. A vivid first-hand account of the pilgrimage can be found in Vincente Rivella's essay accompanying his 1999 exhibit, "Qoyllur Rit'i: In Search of the Lord of The Snow Star" at the W. E. B du Bois Library at the University of Massachusetts, Amherst: http://people.umass.edu/~iespinal/qoyllur_riti/index.html (accessed May 14, 2021).

15. Albrecht et al., "Solastalgia."

BIBLIOGRAPHY

Archival Sources

Alaska and Northern Regions Archive, Rasmussen Library, University of Alaska, Fairbanks, Alaska.
    William O. Field Papers (Field Papers)
American Geographical Society Library Archives. University of Wisconsin–Milwaukee, Milwaukee, Wisconsin.
    American Geographical Society of New York Records, 1723–2010 (AGS Records)
    Field Papers (Field Papers)
Caltech Archives, Pasadena, California.
    Robert P. Sharp Papers (Sharp Papers)
Harvard University Archives, Cambridge, Massachusetts.
    Harvard Mountaineering Club Papers (HMC Papers)
Pennsylvania Historical Society.
    Vaux Family Papers
University of Wisconsin–Milwaukee Libraries, Milwaukee, Wisconsin.
    "Juneau Icefield Research Project" (film).
WMCR. Whyte Museum of the Canadian Rockies, Banff, Alberta.
    Alpine Club of Canada Fonds (ACC Fonds)
    J. Monroe Thorington Papers (Thorington Papers)
    Vaux Family Fonds (Vaux Fonds)
    Arthur Oliver Wheeler Fonds (Wheeler Fonds)

Published Sources

Aalto, K. R. "Rock Stars: Israel Cook Russell (1852–1906)." *GSA Today* 19, no. 2 (February 2009): 14–15.

Agassiz, Louis. *Études Sur Les Glaciers*. Cambridge: Cambridge University Press, 2012 [1840].

Ahlmann, Hans W. "The Contributions of Polar Expeditions to the Science of Glaciology." *Polar Record* 5, no. 37–38 (1949): 324–31.

———. "Foreword." *Journal of Glaciology* 1, no. 1 (1947): 3–4.

———. "Glaciers in Jotunheim and Their Physiography." *Geografiska Annaler* 4 (1922): 1–57.

———. *Land of Fire and Ice*. London: Kegan Paul, Trench, Trubner, 1938.

———. "The Present Climatic Fluctuation." *Geographical Journal* 112, no. 4–6 (1948): 172–76.

———. "Researches on Snow and Ice, 1918–1940." *Geographical Journal* 107, no. 2 (1946): 11–25.

———. "Review: Snow Structure and Ski Fields by C. Seligman." *Geografiska Annaler* 19 (1937): 140–41.

———. "Scientific Results of the Swedish-Norwegian Arctic Expedition in the Summer of 1931, Parts 4–8." *Geografiska Annaler* 15 (1933): 73–216.

Albers, Kate Palmer, and Jordan Bear. "Photography's Time Zones." In *Before-and-After Photography: Histories and Contexts*, edited by Jordan Bear and Kate Palmer Albers, 1–14. London: Bloomsbury Academic Press, 2017.

Albrecht, Glenn, Gina-Maree Sartore, Linda Connor, Nick Higginbotham, Sonia Freeman, Brian Kelly, Helen Stain, Anne Tonna, and Georgia Pollard. "Solastalgia: The Distress Caused by Environmental Change." *Australasian Psychiatry* 15, no. sup1 (2007): S95–S98.

Allen, Garland. *Life Science in the Twentieth Century*. Cambridge: Cambridge University Press, 1978.

Appenzeller, Tim. "The Big Thaw." *National Geographic Magazine* 211, no. 6 (June 2007): 56–71.

Arendt, Anthony A. "Assessing the Status of Alaska's Glaciers." *Science* 332, no. 6033 (May 27, 2011): 1044–45.

Associated Press. "Then and Now: Photographs Document Rapid Melting of World's Glaciers." *CBC News*, April 3, 2017.

Bader, Henri, Robert Haefeli, E. Bucher, E. Neher, J. Eckel, O. Thams, and P. Nigglie. *Snow and Its Metamorphism*. Wilmette, IL: US Army Snow Ice and Permafrost Research Establishment, Corps of Engineers, 1954.

Bailey, Peter. *Leisure and Class in Victorian Britain: Rational Recreation and the Struggle for Control, 1830–1885*. London: Routledge, 1978.

Baird, Patrick D. "A Note on the Commission on Snow and Ice of the International

Association of Scientific Hydrology." *Journal of Glaciology* 3, no. 24 (1958): 253–56.

Baird, Patrick D., and Robert P. Sharp. "Glaciology." *Arctic Research: The Current Status of Research and Some Immediate Problems in the North American Arctic and Subarctic, Special Publication* 2 (1955): 29–40.

Ball, John. "On the Cause and Descent of Glaciers." *Philosophical Magazine and Journal of Science* 40, no. 264 (1870): 1–10.

Balog, James. *Extreme Ice Now: Vanishing Glaciers and a Changing Climate: A Progress Report*. Washington, DC: National Geographic, 2012.

———. *Ice: Portraits of Vanishing Glaciers*. New York: Rizzoli, 2012.

Barasch, Moshe. *Icon: Studies in the History of an Idea*. New York: New York University Press, 1995.

Barnett, Tim P. "Recent Changes in Sea Level and Their Possible Causes." *Climatic Change* 5 (1983): 15–38.

Barthes, Roland. *Camera Lucida: Reflections on Photography*. Translated by Richard Howard. New York: Hill and Wang, 2010.

Beatty, John. "Narrative Possibility and Narrative Explanation." *Studies in the History and Philosophy of Science Part A* 62 (2017): 31–41.

Bella, Leslie. *Parks for Profit*. Montreal: Harvest House, 1987.

Benjamin, Walter. "The Work of Art in the Age of Mechanical Reproduction." In *Illuminations*, edited by Hannah Arendt, 235–36. New York: Schocken, 1935.

Bennett, Timothy Bull, Nancy G. Maynard, Patricia Cochran, Robert Gough, Kathy Lynn, Julie Maldonado, Garrit Voggesser, Susan Wotkyns, and Karen Cozzetto. "Indigenous Peoples, Lands, and Resources." In *Climate Change Impacts in the United States: The Third National Climate Assessment*, edited by Jerry M. Melillo, Therese C. Richmond, and Gary W. Yohe, 297–317. Washington, DC: US Global Change Research Project, 2014.

Benson, Etienne S. *Surroundings: A History of Environments and Environmentalisms*. Chicago: Chicago University Press, 2020.

Benson, Keith R., Ronald Rainger, and Jane Maienschein, eds. *The American Development of Biology*. New Brunswick, NJ: Rutgers University Press, 1991.

Binnema, Theodore, and Melanie Niemi. "'Let the Line Be Drawn Now': Wilderness, Conservation, and the Exclusion of Aboriginal People from Banff National Park in Canada." *Environmental History* 11, no. 4 (2006): 724–50.

Bliss, Michael. *Right Honourable Men*. Toronto: McClelland & Stewart, 1995.

Boime, Albert. *The Magisterial Gaze: Manifest Destiney and American Landscape Painting, c. 1830–1865*. Washington, DC: Smithsonian Institution Press, 1991.

Borges, Jorges Luis. *On Exactitude in Science*. Translated by Andrew Hurley. New York: Penguin, 1998.

Boudia, Soraya. "Observing the Environmental Turn through the Global Environment Monitoring System." In *The Surveillance Imperative: Geosciences during the Cold War and Beyond*, edited by Simone Turchetti and Peder Roberts, 195–235. New York: Palgrave MacMillan, 2014.

Boykoff, Maxwell T. *Who Speaks for the Climate? Making Sense of Media Reporting on Climate Change*. New York: Cambridge University Press, 2011.

Brandt, Allan M. *The Cigarette Century: The Rise, Fall, and Deadly Persistence of the Product That Defined America*. New York: Basic, 2009.

Brglez, Karen. "Surveying Indigenous Space on the Canadian Prairies: The Case of William Wagner." *Prairie History* 6 (Fall 2021): 5–22.

Bridgland, Morris P. "Photographic Surveying in Canada." *Geographical Review* 2, no. 1 (1916): 19–26.

Browne, Janet. "A Science of Empire: British Biogeography Before Darwin." *Revue d'Histoire Des Sciences Humaines* 45, no. 4 (1992): 453–75.

Brückner, Eduard. "Fluctuations of Water Levels in the Caspian Sea, the Black Sea, and the Baltic Sea Relative to Weather." In *Eduard Brückner—The Sources and Consequences of Climate Change and Climate Variability in Historical Times*, edited by Nico Stehr and Hans von Storch, 47–62. Amsterdam, The Netherlands: Kluwer Academic, 2000.

Bruns, Catherine J. "Grieving Okjökull: Discourses on the Ok Glacier Funeral." In *Climate Change and Journalism: Negotiating Rifts of Time*, edited by Henrik Bødker and Hanna E. Marris, 121–35. London: Routledge, 2022.

Brysse, Keynyn, Naomi Oreskes, Jessica O'Reilly, and Michael Oppenheimer. "Climate Change Prediction: Erring on the Side of Least Drama?" *Global Environmental Change* 23, no. 1 (February 2013): 327–37.

Burnett, D. Graham. *Masters of All They Surveyed: Exploration, Geography, and a British El Dorado*. Chicago: University of Chicago Press, 2000.

Callison, Candis. *How Climate Change Comes to Matter: The Communal Life of Facts*. Durham, NC: Duke University Press, 2014.

Camerini, Jane. "Wallace in the Field." *Osiris* 11 (1996): 44–65.

Campbell, Joseph. *The Hero with a Thousand Faces*. Princeton, NJ: Princeton University Press, 1949.

Campbell, Robert. *In Darkest Alaska: Travel and Empire Along the Inside Passage*. Philadelphia: University of Pennsylvania Press, 2007.

Cantor, Geoffrey. "Quakers and Science." In *The Oxford Handbook of Quaker Studies*, edited by Stephen W. Angell and Ben Pink Dandelion, 520–34. Oxford: Oxford University Press, 2013.

Carey, Mark. "The History of Ice: How Glaciers Became an Endangered Species." *Environmental History* 12, no. 3 (2007): 497–527.

———. *In the Shadow of Melting Glaciers: Climate Change and Andean Society*. New York: Oxford University Press, 2010.

———. "The Trouble with Climate Change and National Parks." In *National Parks Beyond the Nation: Global Perspectives on "America's Best Idea,"* edited by Adrian

Howkins, Jared Osri, and Mark Fiege, 258–77. Norman: Oklahoma University Press, 2016.

Carey, Mark, M. Jackson, Alessandro Antonello, and Jaclyn Rushing. "Glaciers, Gender, and Science: A Feminist Glaciological Framework for Global Environmental Change Research." *Progress in Human Geography* 40, no. 6 (2016): 770–93.

Carter, Luther. "Icebergs and Oil Tankers: USGS Glaciologists Are Concerned." *Science* 190, no. 4215 (November 14, 1975): 641–43.

Catton, Theodore. *Inhabited Wilderness: Indians, Eskimos, and National Parks in Alaska*. Albuquerque: University of New Mexico Press, 1997.

Cavell, Edward. *Legacy in Ice: The Vaux Family and the Canadian Alps*. Banff, AB: Whyte Foundation, 1983.

Chakrabarty, Dipesh. "The Climate of History: Four Theses." *Critical Inquiry* 25, no.2 (2009): 197–222.

———. *Provincializing Europe: Postcolonial Thought and Historical Difference*. Princeton, NJ: Princeton University Press, 2000.

Champoux, André, and C. Simon Ommanney. "Evolution of the Illecillewaet Glacier, Glacier National Park, BC, Using Historical Data, Aerial Photography and Satellite Image Analysis." *Annals of Glaciology* 8 (1986): 31–33.

Chu, Pey-Yi. "Mapping Permafrost Country: Creating an Environmental Object in the Soviet Union, 1920s–1940s." *Environmental History* 20, no. 3 (2015): 396–421.

Clements, Philip W. *Science in an Extreme Environment: The 1963 American Mount Everest Expedition*. Pittsburgh, PA: University of Pittsburgh Press, 2018.

Cogley, J. G., R. Hock, L. A. Rasmussen, A. A. Arendt, A. Bauer, R. J. Braithwaite, P. Jansson, et al. "Glossary of Glacier Mass Balance and Related Terms." *IHP-VII Technical Documents in Hydrology* 86 (2011).

Coleman, William. *Biology in the Nineteenth Century: Problems of Form, Function, and Transformation*. Cambridge: Cambridge University Press, 1977.

Colgan, William, W. Tad Pfeffer, H. Rajaram, W. Abdalati, and James Balog. "Monte Carlo Ice Flow Modeling Projects a New Stable Configuration for Columbia Glacier, Alaska, c. 2020." *Cryosphere* 6 (2012): 1395–1409.

Collins, Harry. "Reproducibility of Experiments: Experimenters' Regress, Statistical Uncertainty, and the Reproducibility Imperative." In *Reproducibility: Principles, Problems, Practices, and Prospects*, edited by Harald Atmanspacher and Sabine Maasen, 65–82. Hoboken, NJ: John Wiley & Sons, 2016.

———. "The TEA Set: Tacit Knowledge and Scientific Networks." *Science Studies* 4 (1974): 165–86.

Connor, Cathy, Greg Streveler, Austin S. Post, Daniel Monteith, and Wayne Howell. "The Neoglacial Landscape and Human History of Glacier Bay, Glacier Bay National Park and Preserve, Southeast Alaska, USA." *Holocene* 19, no. 3 (2009): 381–93.

Conway, Howard, L. A. Rasmussen, and Hans-Peter Marshall. "Annual Mass Balance of Blue Glacier, 1955–97." *Geografiska Annaler* 81, no. 4 (1999): 509–20.

Cooper, William S. "The Problem of Glacier Bay, Alaska: A Study of Glacier Variations." *Geographical Review* 27, no. 1 (1937): 37–62.

Cosgrove, Denis. "Images and Imagination in 20th-Century Environmentalism: From
the Sierras to the Poles." *Environment and Planning A* 40, no. 8 (2008): 1862–80.

Cosgrove, Denis, and Veronica della Dora. *High Places: Cultural Geographies of Mountains, Ice and Science*. London: I. B. Taurus, 2009.

Costa, Elio, and Gabriele Scardellato. *Lawrence Grassi: From Piedmont to the Rocky
Mountains*. Toronto: University of Toronto Press, 2015.

Croll, James. "On the Physical Cause of the Motion of Glaciers: Discussion of Moseley's
1869 Paper." *Philosophical Magazine* 37 (1869): 201–6.

Cronon, William. "The Trouble with Wilderness, or Getting Back to the Wrong
Nature." In *Uncommon Ground: Rethinking the Human Place in Nature*, edited by
William Cronon, 69–90. New York: W. W. Norton, 1996.

Cruikshank, Julie. "Are Glaciers 'Good to Think With?': Recognising Indigenous Environmental Knowledge." *Anthropological Forum* 22, no. 3 (2012): 239–50.

————. *Do Glaciers Listen? Local Knowledge, Colonial Encounters, and Social Imagination*. Vancouver: University of British Columbia Press, 2005.

Cruise, David, and Alison Griffiths. *Lords of the Line: The Men Who Built the Canadian Pacific Railroad*. Toronto: Viking Penguin, 1988.

Curl, Herbert Charles, Kenneth Barton, and Lori Harris. "Oil Spill Case Histories,
1967–1991: Summaries of Significant U.S. and International Oil Spills." Seattle: US
National Oceanic and Atmospheric Administration, Hazardous Materials Response
and Assessment Division, 1992.

Darling, Basil S. "Up the Bow and Down the Yoho." *Canadian Alpine Journal* 3 (1911):
157–71.

Daston, Lorraine. "The History of Science and the History of Knowledge." *KNOW* 1,
no. 1 (2017): 131–54.

Daston, Lorraine, and Peter Galison. *Objectivity*. New York: Zone, 2007.

Daston, Lorraine, and Fernando Vidal. *The Moral Authority of Nature*. Chicago: University of Chicago Press, 2004.

Dauenhauer, Nora Marks, and Richard Dauenhauer. *Haa Shuká, Our Ancestors*. Seattle:
University of Washington Press, 1997.

Debarbieux, Bernard, and Gilles Rudaz. *The Mountain: A Political History from the
Enlightenment to the Present*. Chicago: University of Chicago Press, 2015.

De Geer, Gerard. "A Geochronology of the Last 12,000 Years." *Eleventh International
Geological Congress, Stockholm* 1 (1912): 241–53.

Demuth, Bathsheba. *Floating Coast: An Environmental History of the Bering Strait*.
New York: W. W. Norton, 2019.

Demuth, Michael. *Becoming Water: Glaciers in a Warming World*. Victoria, BC: Rocky
Mountain Books, 2012.

Dennis, Michael Aaron. "Earthly Matters: On the Cold War and the Earth Sciences." *Social Studies of Science* 33, no. 5 (2003): 809–19.

Dodds, Klaus. *Geopolitics in Antarctica: Views from the Southern Oceanic Rim*. Chichester, UK: John Wiley and Sons, 1997.

Doel, Ronald E. "Constituting the Postwar Earth Sciences: The Military's Influence on the Environmental Sciences in the USA after 1945." *Social Studies of Science* 33, no. 5 (2003): 635–66.

Doel, Ronald E., Robert Marc Friedman, Julia Lajus, Sverker Sörlin, and Urban Wråkberg. "Strategic Arctic Science: National Interests in Building Natural Knowledge—Interwar Era through the Cold War." *Journal of Historical Geography* 44 (2014): 60–80.

Doel, Ronald E., Kristine Harper, and Matthias Heymann. *Exploring Greenland: Cold War Science and Technology on Ice*. New York: Palgrave MacMillan, 2016.

Doel, Ronald E., Urban Wråkberg, and Suzanne Zeller. "Science, Environment, and the New Arctic." *Journal of Historical Geography* 44 (2014): 2–14.

Donnelly, Peter. "The Invention of Tradition and the (Re)Invention of Mountaineering." In *Method and Methodology in Sport and Cultural History*, edited by K. B. Wamsley, 235–43. Dubuque, IA: Brown & Benchmark, 1995.

Dittmer, Kyle. "Changing Streamflow on Columbia Basin Tribal Lands—Climate Change and Salmon." *Climatic Change* 120 (2013): 627–41.

Dunaway, Finis. "Seeing Global Warming: Contemporary Art and the Fate of the Planet." *Environmental History* 14, no. 1 (2009): 9–31.

———. *Seeing Green: The Use and Abuse of American Environmental Images*. Chicago: University of Chicago Press, 2015.

Earle, Neil. "Three Stories, Two Visions: The West and the Building of the Canadian Pacific Railway in Canadian Culture." *Social Science Journal* 36, no. 2 (1999): 341–52.

Edwards, Paul N. *The Closed World: Computers and the Politics of Discourse in Cold War America*. Cambridge, MA: MIT Press, 1996.

———. "Meteorology as Infrastructural Globalism." *Osiris* 21, no. 1 (2006): 229–50.

———. *A Vast Machine: Computer Models, Climate Data, and the Politics of Global Warming*. Cambridge, MA: MIT Press, 2010.

Endersby, Jim. *Imperial Nature: Joseph Hooker and the Practices of Victorian Science*. Chicago: University of Chicago Press, 2008.

Farber, Paul Lawrence. *Finding Order in Nature: The Naturalist Tradition from Linnaeus to E. O. Wilson*. Baltimore: Johns Hopkins University Press, 2000.

Fay, Charles E. "The Resort of Glacier House, British Columbia." *Appalachia* 9, no. 3–4 (1901): 273–76.

Ferrari, Marco, Elisa Pasqual, and Andrea Bagnato. *A Moving Border: Alpine Cartographies of Climate Change*. New York: Columbia Books on Architecture and the City, 2019.

Field, Frederick. *From Right to Left: An Autobiography*. New York: Lawrence Hill, 1983.

Field, William O. "The Fairweather Range: Mountaineering and Glacier Studies."
Bulletin of the Appalachian Mountain Club 20, no. 4 (1926): 460–72.

———. "Glaciers." *Scientific American* 193, no. 3 (September 1995): 84–95.

———. "The Glaciers of the Northern Part of Prince William Sound, Alaska." *Geographical Review* 22, no. 3 (1932): 361–88.

———. "Glacier Studies in Alaska." *American Alpine Journal* 4, no. 3 (1942): 489–90.

———. "In Search of Mount 'Clearwater.'" *Harvard Mountaineering Bulletin* 1, no. 1
(June 1927): 5–11.

———. "Mountaineering on the Columbia Icefield, 1924." *Bulletin of the Appalachian Mountain Club* 18, no. 2 (1925): 144–54.

———. "Observations on Alaskan Coastal Glaciers in 1935." *Geographical Review* 27,
no. 1 (1937): 63–81.

Field, William O., and Suzanne C. Brown. *With a Camera in My Hands: William O.
Field, Pioneer Glaciologist*. Fairbanks: University of Alaska Press, 2004.

Field, William O., and Maynard M. Miller. "The Juneau Icefield Research Project."
Geographical Review 40, no. 2 (1950): 179–90.

Fisher, Joel E. "Studies of Forbes Dirt Bands." *American Alpine Journal* 6, no. 3 (1947):
309–17.

Fisher, Joel E., Weldon F. Heald, and Maynard M. Miller. "The American Alpine Club
Research Fund, 1946." *American Alpine Journal* 6 (1947): 328–43.

Fleagle, Robert G. *Eyewitness: Evolution of the Atmospheric Sciences*. Boston: American
Meteorological Society, 2001.

Fleming, James Rodger. *Historical Perspectives on Climate Change*. Oxford: Oxford
University Press, 2005.

Flint, Richard Foster. "The American Alpine Club as an Auxiliary in Glacial Research."
American Alpine Journal 7, no. 2 (1949): 305–8.

Flohn, Hermann. "Climatic Change, Ice Sheets and Sea Level." In Proceedings of the
Canberra Symposium on Sea Level, Ice and Climate Change, Canberra, December
1979, edited by Ian Allison, 431–40. Wallingford, UK: International Association of
Hydrological Sciences, 1981.

Flusser, Vilém. *Toward a Philosophy of Photography*. San Francisco: Reaktion, 2000.

Forbes, James David. *Travels through the Alps of Savoy and Other Parts of the Pennine
Chain: With Observations on the Phenomena of Glaciers*. Edinburgh: A. and C.
Black, 1843.

Ford, James D., Nia King, Eranga K. Galappaththi, Tristan Pearce, Graham McDowell,
and Sherilee L. Harper. "The Resilience of Indigenous Peoples to Environmental
Change." *One Earth* 2, no. 6 (2020).

Forel, François A. "Periodic Variations of Alpine Glaciers." *Nature* 1190, no. 46 (August
18, 1892): 386.

Foucault, Michel. "Introduction to Kant's *Anthropology*." In *Introduction to Kant's*

Anthropology, translated by Roberto Nigro and Kate Briggs, 17–124. Los Angeles: Semiotext[e], 2008.

——— . *The Order of Things: An Archaeology of the Human Sciences.* New York: Routledge, 2002.

——— . "What Is Critique?" In *The Politics of Truth,* translated and edited by Lysa Hochroth and Catherine Porter, 41–82. Los Angeles: Semiotext[e], 2007.

Fountain, Andrew G., and Michele A. Funk. "South Cascade Glacier Bibliography." USGS Open-File Report. Tacoma, WA: USGS, Glaciology, 1984.

Fowkes, Maja, and Reuben Fowkes. *Art and Climate Change.* London: Thames and Hudson, 2022.

Fox, Stephen R. *The American Conservation Movement: John Muir and His Legacy.* Madison: University of Wisconsin Press, 1985.

Francis, Daniel. *National Dreams: Myth, Memory, and Canadian History.* Vancouver, BC: Arsenal Pulp, 1997.

Frank, Susi K., and Kjetil A. Jakobsen. *Arctic Archives: Ice, Memory and Entropy.* Bielefeld, Germany: Transcript, 2019.

Gagné, Karine. *Caring for Glaciers: Land, Animals, and Humanity in the Himalayas.* Seattle: University of Washington Press, 2018.

Garrard, Rodney, and Mark Carey. "Beyond Images of Melting Ice: Hidden Histories of People, Place, and Time in Repeat Photography of Glaciers. In *Before-and-After Photography: Histories and Contexts,* edited by Jordan Bear and Kate Palmer Albers, 101–22. London: Bloomsbury Academic Press, 2017.

Gieryn, Thomas F. "Boundary-Work and the Demarcation of Science from Non-Science: Strains and Interests in Professional Ideologies of Scientists." *American Sociological Review* 48, no. 6 (1983): 781–95.

Gilbert, Grove Karl. *Glaciers and Glaciation, Harriman Alaska Expedition.* Vol. 3. New York: Doubleday, Page, 1910.

"Glacier Section." *Canadian Alpine Journal* 22 (1932): 21–22.

Goetzmann, William H. *Exploration and Empire: The Explorer and the Scientist in the Winning of the American West.* New York: History Book Club, 1993.

Goetzmann, William H., and Kay Sloan. *Looking Far North: The Harriman Expedition to Alaska, 1899.* Princeton, NJ: Princeton University Press, 1982.

Goldthwait, Richard P. "Seismic Sounding on South Crillon and Klooch Glaciers." *Geographical Review* 87, no. 6 (1936): 496–511.

Grant, Ulysses S., and Daniel F. Higgins. "Coastal Glaciers of Prince William Sound and Kenai Peninsula, Alaska." *U.S. Geological Survey Bulletin* 526 (1913): 7–40.

Green, William Spotswood. *Among the Selkirk Glaciers: Being the Account of a Rough Survey in the Rocky Mountain Regions of British Columbia.* London: Macmillan, 1890.

Greene, Mott. *Alfred Wegener: Science, Exploration, and the Theory of Continental Drift.* Baltimore: Johns Hopkins University Press, 2015.

Haefeli, Robert. "The Development of Snow and Glacier Research in Switzerland." *Journal of Glaciology* 1, no. 4 (1948): 192–201.

Ham, George H. *Reminiscences of a Raconteur, Between the '40s and the '20s*. Toronto: Musson, 1921. Project Gutenberg ed.

Hamblin, Jacob Darwin. *Arming Mother Nature: The Birth of Catastrophic Environmentalism*. Oxford: Oxford University Press, 2013.

Handy, Dora Keen. "First Exploration of the Harvard Glacier." *Bulletin of the American Geographical Society* 47, no. 2 (1915): 117–19.

Hansen, Peter H. "Albert Smith, the Alpine Club, and the Invention of Modern Mountaineering in Mid-Victorian Britain." *Journal of British Studies* 34, no. 3 (1995): 300–324.

———. *The Summits of Modern Man: Mountaineering After the Enlightenment*. Cambridge, MA: Harvard University Press, 2013.

Haraway, Donna J. *Primate Visions: Gender, Race, and Nature in the World of Modern Science*. New York: Routledge, 1989.

Harding, Sandra. *The Science Question in Feminism*. Ithaca, NY: Cornell University Press, 1986.

Harris, Cole. *A Bounded Land: Reflections on Settler Colonialism*. Vancouver: University of British Columbia Press, 2020.

Hart, Edward J. *The Place of Bows: Exploring the Heritage of the Banff-Bow Valley, Part 1*. Banff, AB: EJH Literary, 1999.

Heggie, Vanessa. *Higher and Colder: A History of Extreme Physiology and Exploration*. Chicago: University of Chicago Press, 2019.

Heusser, Calvin J. *Juneau Icefield Research Project: 1948–1958*. Oxford: Elsevier, 2007.

Hevly, Bruce. "The Heroic Science of Glacial Motion." *Osiris* 11 (1996): 66–86.

Hindle, Brooke. "The Quaker Background and Science in Colonial Philadelphia." *Isis* 46, no. 3 (1955): 243–50.

Houghton, John T., Geoff J. Jenkins, and J. Ephramus, eds. *Climate Change: The IPCC Scientific Assessment*. Cambridge: Cambridge University Press, 1990.

Howe, Joshua P. *Behind the Curve: Science and the Politics of Global Warming*. Seattle: University of Washington Press, 2014.

———, ed. *Making Climate Change History: Documents from Global Warming's Past*. Seattle: University of Washington Press, 2017.

———. "This Is Nature; This Is Un-Nature: Reading the Keeling Curve." *Environmental History* 20, no. 2 (2015): 286–93.

Howkins, Adrian. "Reluctant Collaborators: Argentina and Chile in Antarctica during the IGY." *Journal of the History of Geography* 34, no. 4 (2008): 596–617.

———. "Science, Environment, and Sovereignty: The International Geophysical Year in the Antarctic Peninsula Region." In *Globalizing Polar Science: Reconsidering the International Polar and Geophysical Years*, edited by Roger D. Launius, James Rodger Fleming, and David H. deVorkin, 245–64. New York: Palgrave Macmillan, 2010.

———. "The Significance of the Frontier in Antarctic History: How the U.S. West Has Shaped the Geopolitics of the Far South." *Polar Journal* 3, no. 1 (2003): 9–30.

Hubbard, Gardiner, Marcus Baker, and Willard D. Johnson. "'Memorandum of Instructions': An Expedition to Mount St. Elias, Alaska." *National Geographic Magazine* 3 (1891): 53–204.

Hubley, Richard C. "Recent Growth of Glaciers in the Pacific Northwest." *American Alpine Journal* 10, no. 1 (1956): 162–63.

Hughes, Terence J. "John H. Mercer (1922–1987)." *Journal of Glaciology* 34, no. 16 (1988): 136–38.

Hulme, Mike. *Why We Disagree about Climate Change: Understanding Controversy, Inaction, and Opportunity*. Cambridge: Cambridge University Press, 2009.

Imbrie, John, and Katherine Palmer Imbrie. *Ice Ages: Solving the Mystery*. Cambridge, MA: Harvard University Press, 1986.

Inkpen, Dani K. "Ever Higher: The Mountain Cryosphere." In *Ice Humanities*, edited by Klaus Dodds and Sverker Sörlin, 72–88. Manchester, England: Manchester University Press, 2022.

———. "Frozen Icons: The Science and Politics of Repeat Glacier Photographs, 1887–2010." PhD diss., Harvard University, 2018.

———. "Of Ice and Men: The Evolving Role of the Camera in Twentieth-Century Glacier Study." *Environmental History* 27, no. 3 (July 2022): 547–60.

———. "The Scientific Life in the Alpine: Recreation and Moral Life in the Field." *Isis* 109, no. 3 (September 2018): 525–37.

Inkpen, S. Andrew. "Are Humans Disturbing Conditions in Ecology?" *Biology and Philosophy* 32, no. 1 (2017): 51–71.

———. "Demarcating Nature, Defining Ecology: Creating a Rationale for the Study of Nature's "Primitive Conditions." *Perspectives on Science* 25, no. 3 (2017): 355–92.

———. "Like Hercules and the Hydra: Trade-Offs and Strategies in Ecological Model-Building and Experimental Design." *Studies in History and Philosophy of Biological and Biomedical Sciences* 57 (2016): 34–43.

IPCC (Intergovernmental Panel on Climate Change). "Climate Change 1995: IPCC Second Assessment, A Report of the Intergovernmental Panel on Climate Change." Cambridge: Cambridge University Press, 1996.

Isserman, Maurice. *Continental Divide: A History of American Mountaineering*. New York: W. W. Norton, 2015.

Ives, Katie. "Vanishing Uplands." *Alpinist* 60 (Winter 2017–18): 13.

Jackson, M. "Glaciers and Climate Change: Narratives of Ruined Futures." *Wiley Interdisciplinary Reviews: Climate Change* 6, no. 5 (2015): 479–92.

———. *The Secret Lives of Glaciers*. Brattleboro, VT: Greenwriters, 2019.

Jacoby, Karl. *Crimes Against Nature: Squatters, Poachers, Thieves, and the Hidden History of American Conservation*. Berkeley: University of California Press, 2001.

Jardine, Nicholas, James A. Secord, and Emma C. Spary, eds. *Cultures of Natural History*. Cambridge: Cambridge University Press, 1996.

Jasanoff, Sheila. "A New Climate for Society." *Theory, Culture, Society* 27 (2010): 233–53.

Jones, Majorie G. "Bowling Along: Early Travel Adventures of Mary Morris Vaux." *Quaker History* 100, no. 1 (2011): 22–39.

———. *The Life and Times of Mary Vaux Walcott*. Atglen, PA: Schiffer, 2015.

Jones-Imhotep, Edward. *The Unreliable Nation: Hostile Nature and Technological Failure in the Cold War*. Cambridge, MA: MIT Press, 2017.

Jordanova, Ludmilla. *Sexual Visions: Images of Gender in Science and Medicine between the Eighteenth and Twentieth Centuries*. Madison: University of Wisconsin Press, 1989.

Kintisch, Eli. "Meltdown." *Science* 355, no. 6327 (February 24, 2017): 788–91.

Kirwan, Laurence P., Carl Mannerfelt, Carl-Gustaf Rossby, and Valter Schytt. "Glaciers and Climatology: Hans W. Ahlmann's Contribution." *Geografiska Annaler* 31 (1949): 11–13.

Knight, Peter G. *Glaciers: Nature and Culture*. London: Reaktion, 2019.

Kohler, Robert E. *All Creatures: Naturalists, Collectors, and Biodiversity, 1850–1950*. Princeton, NJ: Princeton University Press, 2006.

———. *Landscapes and Labscapes: Exploring the Lab-Field Border in Biology*. Chicago: University of Chicago Press, 2002.

Kohler, Robert E., and Kathryn M. Olesko. *Clio Meets Science: The Challenges of History*. Chicago: University of Chicago Press, 2012.

Kohler, Robert E., and Jeremy Vetter. "The Field." In *A Companion to the History of Science* edited by Bernard Lightman, 282–95. Hoboken, NJ: John Wiley & Sons, 2016.

Korsmo, Fae. "The Early Cold War and U.S. Arctic Research." In *Extremes: Oceanography's Adventures at the Poles*, edited by Katherine Anderson, Keith R. Benson, and Helen Rozwadowski, 173–200. Sagamore Beach, MA: Watson, 2007.

Kuklick, Henricka, and Robert E. Kohler, eds. *Science in the Field. Osiris* 11, 2nd ser. Chicago: University of Chicago University Press, 1996.

LaChapelle, Edward R. "Assessing Glacier Mass Budgets by Reconnaissance Aerial Photography." *Journal of Glaciology* 4, no. 33 (1962): 290–97.

Lackenbauer, P. Whitney, and Matthew Farish. "The Cold War on Canadian Soil: Militarizing a Northern Environment." *Environmental History* 12, no. 4 (2007): 902–50.

Ladurie, Emmanuel Le Roy. *Times of Feast, Times of Famine: A History of Climate Since the Year 1000*. New York: Doubleday, 1971.

Lamb, William Kaye. *History of the Canadian Pacific Railway*. New York: Macmillan, 1977.

Langway, Chester C. *History of Early Polar Ice Cores*. Hanover, NH: US Army Cold Regions Research and Engineering Laboratory, 2008.

Laslett, Barbara, Sally Gregory Kohlstedt, Helen Longino, and Evelynn Hammonds, eds. *Gender and Scientific Authority*. Chicago: University of Chicago Press, 1996.

Latour, Bruno. "Visualization and Cognition: Drawing Things Together." *Knowledge and Society: Studies in the Sociology of Culture and Present* 6 (1986): 1–40.

———. "Why Has Critique Run Out of Steam? From Matters of Fact to Matters of Concern." *Critical Inquiry* 30, no. 2 (2004): 225–48.

Liboiron, Max. *Pollution Is Colonialism.* Durham, NC: Duke University Press, 2021.

Livingstone, David. *Putting Science in Its Place: Geographies of Scientific Knowledge.* Chicago: University of Chicago Press, 2003.

Longino, Helen. "Evidence and Hypothesis: An Analysis of Evidential Relations." *Philosophy of Science* 46, no. 1 (1979): 35–56.

Loo, Tina. *States of Nature: Conserving Canada's Wildlife in the Twentieth Century.* Vancouver: University of British Columbia Press, 2006.

Lüdecke, Cornelia. "Lifting the Veil: The Circumstances That Caused Alfred Wegener's Death on the Greenland Icecap, 1930." *Polar Record* 36, no. 197 (2000): 139–54.

Luthcke, Scott B., T. J. Sabaka, B. D. Loomis, A. A. Arendt, J. J. McCarthy, and J. Camp. "Antarctica, Greenland and Gulf of Alaska Land-Ice Evolution from an Iterated GRACE Global Mascon Solution." *Journal of Glaciology* 59, no. 216 (2013): 613–31.

MacBeth, Roderick George. *The Romance of the Canadian Pacific Railway.* Toronto: Ryerson, 1924.

MacDonald, Robert. "Challenges and Accomplishments: A Celebration of the Arctic Institute of North America." *Arctic* 58, no. 4 (2005): 440–51.

Macfarlane, Robert. *Mountains of the Mind: How Desolate and Forbidding Heights Were Transformed into Experiences of Indomitable Spirit.* New York: Pantheon, 2003.

MacLaren, Ian S., Eric Higgs, and Gabrielle Zezulka-Mailloux. *Mapper of Mountains: M. P. Bridgland in the Canadian Rockies, 1902–1930.* Edmonton: University of Alberta Press, 2005.

Malm, Andreas, and Alf Hornborg. "The Geology of Mankind? A Critique of the Anthropocene Narrative." *Anthropocene Review* 1, no. 1 (2014): 62–69.

Manley, Gordon. "Some Recent Contributions to the Study of Climatic Change." *Quarterly Journal of the Royal Meteorological Society* 70, no. 305 (1944): 197–219.

Marsh, John. "The Rocky and Selkirk Mountains and the Swiss Connection." *Annals of Tourism Research* 12, no. 3 (1985): 417–33.

Martin, Lawrence. "Alaskan Glaciers in Relation to Life." *Bulletin of the American Geographical Society* 45, no. 11 (1919): 801–18.

———. "Glacial Scenery in Alaska." *Bulletin of the American Geographical Society* 47, no. 3 (1915): 172–74.

Martin-Nielsen, Janet. *Eismitte in the Scientific Imagination: Knowledge and Politics at the Center of Greenland.* New York: Palgrave MacMillan, 2013.

———. "The Other Cold War: The United States and Greenland's Ice Sheet Environment, 1949–1966." *Journal of Historical Geography* 68, no. 1 (2012): 68–80.

Marzeion, Ben J., Graham Cogley, Kristin Ritcher, and David Parkes. "Attribution of

Global Glacier Mass Loss to Anthropogenic and Natural Causes." *Science* 345, no. 6199 (August 22, 2014): 919–21.

Mason, Courtney W. "The Construction of Banff as a 'Natural' Environment: Sporting Festivals, Tourism, and Representations of Aboriginal Peoples." *Journal of Sport History* 35, no. 2 (2008): 221–39.

Mathews, William H. "Glacier Study for the Mountaineer." *Canadian Alpine Journal* 36 (1953): 161–67.

Matthes, François. "Max Harrison Demorest." *EOS Transactions of the American Geophysical Union* 24, no. 1 (1943): 121–16.

Matthews, William. "Mechanical Properties of Ice and Their Relation to Glacier Motion." *Nature* 1, no. 21 (1870): 534–35.

McNeill, John R. *Something New Under the Sun: An Environmental History of the Twentieth Century*. New York: W. W. Norton, 2000.

Meier, Mark F. "Contributions of Small Glaciers to Global Sea Level Rise." *Science* 226, no. 4681 (December 21, 1984): 1418–21.

———. "Ice, Climate, and Sea Level: Do We Know What Is Happening?" In *Ice in the Climate System*, edited by William Richard Peltier, 141–60. Berlin: Springer-Verlag, 1993.

———. "Mass Budget of South Cascade Glacier, 1957–60." In *United States Geological Survey Professional Paper* 424-B: 206–11. Tacoma, WA: USGS, 1961.

———. "Research on the South Cascade Glacier." *Mountaineers* 51, no. 4 (1958): 40–47.

———. "Role of Land Ice in Present and Future Sea-Level Change." In *Studies in Geophysics,* edited by Roger Revelle, 171–84. Washington, DC: National Academies Press, 1990.

———. "Snow and Ice in a Changing Hydrological World." *Hydrological Sciences Journal* 28, no. 1 (1983): 3–22.

Meier, Mark F., Mark B. Dyurgerov, Ursula K. Rick, Shad O'Neel, W. Tad Pfeffer, Robert S. Anderson, Suzanne P. Anderson, and Andrey Glazovsky. "Glaciers Dominate Eustatic Sea-Level Rise in the 21st Century." *Science* 317, no. 5841 (August 24, 2007): 1064–67.

Meier, Mark F., and John M. Wahr. "Sea Level Is Rising: Do We Know Why?" *Proceedings of the National Academy of Sciences* 99, no. 10 (2002): 6524–26.

Menzies, Archibald. *The Travel Diaries of Archibald Menzies, 1793–1794*. Fairbanks: University of Alaska Press, 1993.

Mercer, John. "West Antarctic Ice Sheet and CO_2 Greenhouse Effect: A Threat of Disaster." *Nature* 271 (January 26, 1978): 321–25.

Merton, Robert K. *The Sociology of Science: Theoretical and Empirical Investigations*. Chicago: University of Chicago Press, 1973.

Miller, Maynard M. "Mount Bertha: Fairweather Range, Alaska—1940." *Mountaineers* 33, no. 1 (December 1940): 48–52.

———. "Twenty-Five Explorers Probe an Alaskan Mystery . . . Vanishing Glaciers." *Science Illustrated* 4, no. 1 (1949): 21–27.

Moseley, Henry. "On the Descent of Glaciers." *Proceedings of the Royal Society of London* 7 (1856): 333–42.

Mouginot, J., E. Rignot, and B. Scheuchl. "Sustained Increase in Ice Discharge from the Amundsen Sea Embayment, West Antarctica, from 1973 to 2013." *Geophysical Research Letters* 41, no. 5 (2014): 1576–84.

"Mountaineering Club of Revelstoke." *Canadian Alpine Journal* 2, no. 1 (1909): 106–7.

Muir, John. "Alaska." *American Geologist* 11, no. 5 (1893): 295–99.

———. *Alaska via Northern Pacific R.R.* St. Paul, MN: Northern Pacific Railroad, 1891.

———. "The Discovery of Glacier Bay." *Century Magazine* 50, no. 2 (1895): 234–37.

Murakami, Haruki. *What I Talk About When I Talk About Running.* New York: Knopf, 2008.

Nash, Roderick. *Wilderness and the American Mind.* 5th ed. New Haven, CT: Yale University Press, 2015 [1967].

National Research Council. "Carbon Dioxide and Climate: A Scientific Assessment." Report of an Ad Hoc Study Group on Carbon Dioxide and Climate. Climate Research Board Assembly of Mathematical and Physical Sciences National Research Council, Woods Hole, Massachusetts, July 23, 1979.

National Research Council, Ad Hoc Committee on the Relationship between Land Ice and Sea Level, Committee on Glaciology, Polar Research Board, and Commission on Physical Sciences, Mathematics, and Resources. Glaciers, Ice Sheets, and Sea Level: Effect of a CO_2-Induced Climatic Change Report of a Workshop Held in Seattle, Washington, September 13–15, 1984. Washington, DC: National Academy Press, 1985.

Nicholson, Marjorie Hope. *Mountain Gloom and Mountain Glory: The Development of the Aesthetics of the Infinite.* Seattle: University of Washington Press, 1997.

Numbers, Ronald. "George Frederick Wright: From Christian Darwinist to Fundamentalist." *Isis* 79, no. 4 (1988): 624–45.

Nye, John F. "The Flow of Glaciers and Ice-Sheets as a Problem in Plasticity." *Proceedings of the Royal Society of London* A207, no. 1091 (1951): 554–72.

———. "The Mechanics of Glacier Flow." *Journal of Glaciology* 2, no. 12 (1952): 82–93.

Odell, Noel. "Recent Glaciological Work." *Polar Record* 45 (1945): 272–76.

Odishaw, Hugh. "International Geophysical Year." *Science* 129, no. 3340 (January 2, 1958): 14–25.

Oelshlaeger, Max. *The Idea of Wilderness.* New Haven, CT: Yale University Press, 1991.

Ohmura, Atsumu. "Completing the World Glacier Inventory." *Annals of Glaciology* 50, no. 53 (2009): 144–48.

Oppenheimer, Michael, Naomi Oreskes, Dale Jamieson, Keynyn Brysse, Jessica O'Reilly, Matthew Shindell, and Milena Wazeck. *Discerning Experts: The Practices of Scientific Assessment for Environmental Policy.* Chicago: University of Chicago Press, 2019.

O'Reilly, Jessica. *The Technocratic Antarctic: An Ethnography of Scientific Expertise and Environmental Governance*. Ithaca, NY: Cornell University Press, 2017.

O'Reilly, Jessica, Naomi Oreskes, and Michael Oppenheimer. "A Rapid Disintegration of Projections: The West Antarctic Ice Sheet and the Intergovernmental Panel on Climate Change." *Social Studies of Science* 42, no. 5 (2012): 709–31.

Oreskes, Naomi. "Beyond the Ivory Tower: The Scientific Consensus on Climate Change." *Science* 306, no. 5702 (December 3, 2004): 1686–87.

———. "From Scaling to Simulation." In *Science Without Laws: Model Systems, Cases, Exemplary Narratives*, edited by Angela Creager, Elizabeth Lunbeck, and M. Norton Wise, 93–124. Durham, NC: Duke University Press, 2007.

———. "Objectivity or Heroism? On the Invisibility of Women in Science." *Osiris* 11 (1996): 87–113.

———. "Of Mavericks and Mules." Paper presented at AAAS Annual Meeting, Washington, DC, February 19, 2011.

———. *The Rejection of Continental Drift: Theory and Method in American Earth Science*. New York: Oxford University Press, 1997.

———. *Science on a Mission: How Military Funding Shaped What We Do and Don't Know about the Ocean*. Chicago: University of Chicago Press, 2021.

Oreskes, Naomi, and Erik M. Conway. *Merchants of Doubt: How a Handful of Scientists Obscured the Truth on Issues from Tobacco Smoke to Global Warming*. London: Bloomsbury, 2010.

Oreskes, Naomi, and Ronald Rainger. "Science and Security before the Atomic Bomb: The Loyalty Case of Harald U. Sverdrup." *Studies in the History and Philosophy of Modern Physics* 31B (2000): 309–69.

Orlove, Ben, Ellen Wiegant, and Brian H. Luckman, eds. *Darkening Peaks: Glacier Retreat, Science, and Society*. Berkeley: University of California Press, 2008.

Orlowski, Jeff, dir. *Chasing Ice*. Documentary, 2013. Submarine Deluxe.

Otter, Andy Albert den. *The Philosophy of Railways: The Transcontinental Railway Idea in British North America*. Toronto: University of Toronto Press, 1997.

Pan, Linda, Evelyn M. Powell, Konstantin Latychev, Jerry X. Mitrovica, Jessica R. Creveling, Natalya Gomez, Mark J. Hoggard, and Peter U. Clark. "Rapid Postglacial Rebound Amplifies Global Sea Level Rise Following West Antarctic Ice Sheet Collapse." *Science Advances* 7, no. 18 (2021): eabf7877.

Parker, Elizabeth. "The Alpine Club of Canada." *Canadian Alpine Journal* 1 (1907): 7.

Patterson, Stan. "Glaciological Research in Western Canada in 1959: Expedition to the Athabaska Glacier." *Canadian Alpine Journal* 43 (1960): 99–103.

Pelto, Mauri. "From a Glacier's Perspective: Glacier Change in a World of Climate Change," September 13, 2015.

Peterson, Thomas, William M. Connolley, and John Fleck. "The Myth of the 1970s Global Cooling Scientific Consensus." *Bulletin of the American Meteorological Society* 89, no. 9 (2008): 1325–37.

Pfeffer, W. Tad. "Glaciology Needs to Come Out of the Ivory Tower." *Earth* 57, no. 11 (2012): 8–10.

———. "Land Ice and Sea Level Rise: A Thirty Year Perspective." *Oceanography* 24, no. 2 (2011): 94–111.

———. "Obituary: Austin Post, 1922–2013." *ICE* 161, no. 1 (2013): 28–33.

———. *The Opening of a New Landscape: Columbia Glacier at Mid-Retreat*. Washington, DC: American Geophysical Union, 2007.

Piper, Liza. "Introduction: The History of Circumpolar Science and Technology." *Scientia Canadensis* 33, no. 2 (2010): 1–9.

Post, Austin. "Alaskan Glaciers: Recent Observations in Respect to the Earthquake-Advance Theory." *Science* 148, no. 3668 (1965): 366–68.

———. "Distribution of Surging Glaciers in Western North America." *Journal of Glaciology* 8, no. 53 (1969): 229–40.

———. "The Exceptional Advances of the Muldrow, Black Rapids, and Sustina Glaciers." *Journal of Geophysical Researches* 65, no. 11 (1960): 3703–12.

Post, Austin, and Edward R. LaChapelle. *Glacier Ice*. Rev. ed. Toronto: University of Toronto Press in association with the International Glaciological Society, 2000.

Powell, Richard C. "Lines of Possession? The Anxious Constitution of a Polar Geopolitics." *Polar Geography* 29 (2010): 74–77.

———. "Science, Sovereignty and Nation: Canada and the Legacy of the International Geophysical Year, 1957–1958." *Journal of Historical Geography* 34 (2008): 618–38.

Powell, Richard C., and Klaus Dodds, eds. *Polar Geopolitics? Knowledge, Resources, and Legal Regimes*. Cheltenham, England: Edward Elgar, 2014.

Pringle, Allan. "William Cornelius van Horne: Art Director, Canadian Pacific Railway." *Journal of Canadian Art History* 8, no. 1 (1984): 50–79.

Pritchard, Sara B. *Confluence: The Nature of Technology and the Remaking of the Rhône*. Cambridge, MA: Harvard University Press, 2011.

"Proceedings of the Moscow Symposium, August 1971." *International Association of Hydrological Sciences* 104 (1975): 185–96.

Proctor, Robert N. *Golden Holocaust: Origins of the Cigarette Catastrophe and the Case for Abolition*. Berkeley: University of California Press, 2011.

Putnam, William Lowell. *The Great Glacier and Its House: The Story of the First Center of Alpinism in North America, 1885–1925*. New York: American Alpine Club, 1982.

———. *Green Cognac: The Education of a Mountain Fighter*. New York: American Alpine Club, 1991.

Qiu, Jane. "Ice on the Run." *Science* 358, no. 6367 (December 1, 2017): 1120–23.

Radok, Uwe. "The International Commission on Snow and Ice (ICSI) and Its Precursors, 1894–1994." *Hydrological Sciences Journal* 42, no. 2 (1997): 131–40.

Reichwein, PearlAnn. *Climber's Paradise: Making Canada's National Parks, 1906–1974*. Edmonton: University of Alberta Press, 2014.

Reid, Harry Fielding. *Glacier Bay and Its Glaciers*. United States Geological Survey,

16th Annual Report, 1894–95. Washington, DC: US Geological Survey, 1896.

———. "Glaciers and Geophysics." *EOS Transactions of the American Geophysical Union* 14, no. 1 (1933): 28–30.

———. "The Variations of Glaciers, XII." *Journal of Geology* 16, no. 1 (1908): 46–55.

Reidy, Michael S. "Evolutionary Naturalism on High: The X-Club Sequester the Alps." In *Victorian Scientific Naturalism: Community, Identity, Continuity*, edited by Gowan Dawson and Bernard Lightman, 54–78. Chicago: University of Chicago Press, 2014.

———. "John Tyndall's Vertical Physics: From Rock Quarries to Icy Peaks." *Physics in Perspective* 12, no. 2 (2010): 122–45.

———. "Mountaineering, Masculinity, and the Male Body in Mid-Victorian Britain." *Osiris* 30, no. 1 (2015): 158–81.

Roberts, Peder. "Walter Wood and the Legacies of Science and Alpinism in the St. Elias Mountains." In *Rethinking Geographical Explorations in Extreme Environments: From the Arctic to the Mountaintops*, edited by Marco Armiero, Roberta Biasillo, and Stefan Morosini, 90–105. London: Routledge, 2022.

Robertson, Alan Bruce. "The First Ascent of Mount Vancouver." *Canadian Alpine Journal* 33 (1950): 1–38.

Robin, Gordon de Quetteville. "Review of *Sea Level, Ice, and Climatic Change*, edited by Ian Allison, International Association of the Hydrological Sciences, 1981." *Quarterly Journal of the Royal Meteorological Society* 109, no. 460 (April 1983): 439–40.

Robinson, Zac. *Conrad Kain: Letters from a Wandering Mountain Guide, 1906–1933*. Edmonton: University of Alberta Press, 2014.

———. "Storming the Heights: Canadian Frontier Nationalism and the Making of Manhood in the Conquest of Mount Robson, 1906–13." *International Journal of the History of Sport* 22, no. 3 (2005): 415–33.

Robinson, Zac, and Stephen Slemon. "The Shining Mountains." *Alpinist Magazine* 50 (2015): 115–24.

Rossiter, Margaret. *Women Scientists in America: Struggles and Strategies to 1940*. Baltimore, MD: Johns Hopkins University Press, 1982.

Rudiak-Gould, Peter. "'We Have Seen It With Our Own Eyes': Why We Disagree About Climate Change Visibility." *Weather, Climate, and Society* 5, no. 2 (2013): 120–32.

Rudwick, Martin. "The Emergence of a Visual Language for Geology, 1760–1840." *History of Science* 14 (1976): 149–95.

Rumore, Gina Maria. "A Natural Laboratory, a National Monument: Carving Out a Place for Science in Glacier Bay, Alaska, 1879–1959." PhD diss., University of Minnesota, 2009.

Russell, Israel Cook. "Climatic Changes Indicated by the Glaciers of North America." *American Geologist* 9 (May 1892): 322–36.

Ryan, James, and Simon Naylor. *New Spaces of Exploration: Geographies of Discovery in the Twentieth Century.* London: Bloomsbury, 2010.

Sanford, Robert William. *Our Vanishing Glaciers: The Snows of Yesteryear and the Future Climate of the Mountain West.* Victoria, BC: Rocky Mountain Books, 2017.

Sapolsky, Harvey M. *Science and the Navy: The History of the Office of Naval Research.* Princeton, NJ: Princeton University Press, 1990.

Savage, Neil. "Climatology on Thin Ice." *Nature* 520, no. 7547 (2015): 395–97.

Schama, Simon. *Landscape and Memory.* London: Harper Perennial, 1995.

Schiebinger, Londa. *Nature's Body: Gender in the Making of Modern Science.* Baltimore: Johns Hopkins University Press, 1993.

Schmidt, Gavin, and Joshua Wolfe. *Climate Change: Picturing the Science.* New York: Wiley, 2009.

Schneider, Stephen. *Science as a Contact Sport: Inside the Battle to Save Earth's Climate.* Washington, DC: National Geographic, 2009.

Schultz, Susan. "The Debate Over Multiple Glaciation in the United States: T. C. Chamberlain and G. F. Wright, 1889–1894." *Earth Sciences History* 2, no. 2 (1983): 122–29.

Schytt, Valter. "A. (S.) Post and E. R. LaChapelle: Glacier Ice." *Journal of Glaciology* 12, no. 65 (1971): 324–25.

Scott, Chic. *Pushing the Limits: The Story of Canadian Mountaineering.* Victoria, BC: Rocky Mountain Books, 2010.

Scurlock, John. "Austin Post: Legendary Chronicler of Glaciers." *Northwest Mountaineering Journal* 4 (2004).

Sfraga, Michael. *Bradford Washburn: A Life of Exploration.* Corvallis: Oregon State University Press, 2004.

Shairp, John Campbell, Peter Guthrie Tait, and Anthony Adams-Reilly, eds. *Life and Letters of James David Forbes.* London: Macmillan, 1873.

Shapin, Steven. *Never Pure: Historical Studies of Science As If It Was Produced by People with Bodies, Situated in Time, Space, Culture, and Society, and Struggling for Credibility and Authority.* Baltimore: Johns Hopkins University Press, 2010.

———. "Placing the View from Nowhere: Historical and Sociological Problems in the Location of Science." *Transactions of the Institute of British Geographers NS* 21, no. 1 (1998): 5–12.

Sharp, Robert P. "Accumulation and Ablation on the Seward-Malaspina Glacier System, Canada-Alaska." *Bulletin of the Geological Society of America* 62, no. 7 (July 1951): 725–44.

Sheppard, Stephen R. J. *Visualizing Climate Change.* London: Routledge, 2012.

Shumskii, Petr A. "The Energy of Glaciation and the Life of Glaciers." U.S. Snow, Ice, and Permafrost Research Establishment, 1950.

———. *Principles of Structural Glaciology: The Petrography of Fresh-Water Ice as a Method of Glaciological Investigation.* Translated by David Kraus. New York: Dover, 1964.

Silverman, Kaja. *The Miracle of Analogy, or The History of Photography, Part 1*. Stanford: Stanford University Press, 2018.

Skidmore, Collen (editor). *This Wild Spirit: Women in the Rocky Mountains of Canada*. Edmonton: University of Alberta Press, 2016.

Smart, David. *Paul Preuss, Lord of the Abyss: Life and Death at the Birth of Free-Climbing*. Victoria, BC: Rocky Mountain Books, 2019.

Snow, John. *These Mountains Are Our Sacred Places*. Toronto: Samuel Stevens, 1977.

Snyder, Elisabeth F. *Bibliography of Glacier Studies by the U.S. Geological Survey*. Washington, DC: US Department of the Interior, 1996.

Snyder, Joel. "Res Ipsa Loquitur." In *Things That Talk: Object Lessons from Art and Science*, edited by Lorraine Daston, 195–222. New York: Zone, 2004.

Sontag, Susan. *On Photography*. New York: Picador, 1973.

Sörlin, Sverker. "The Anxieties of a Science Diplomat: Field Coproduction of Climate Knowledge and the Rise and Fall of Hans Ahlmann's 'Polar Warming.'" *Osiris* 26 (2011): 66–68.

———. "Can Glaciers Speak?: The Political Aesthetics of Vo/Ice." In *New in Methodological Challenges in Nature-Culture and Environmental History Research*, edited by Jocelyn Thorpe, Stephanie Rutherford, and L. Anders Sandberg, 13–30. New York: Routledge, 2017.

———. "Cryo-History: Narratives of Ice and the Emerging Arctic Humanities." In *The New Arctic*, edited by Birgitta Evengård, Joan Nymand Larsen, and Øyvind Paasche, 327–39. Cham, Switzerland: Springer International, 2015.

———. "The Global Warming That Did Not Happen: Historicizing Glaciology and Climate Change." In *Nature's End: History and the Environment*, edited by Sverker Sörlin and Paul Warde, 93–114. London: Palgrave MacMillan, 2009.

———. "A Microgeography of Authority: Glaciology and Climate Change at the Tarfala Station, 1945–1980." In *Understanding Field Science Institutions*, edited by Patience Schell et al., 255–85. London: Science History, 2018.

———. "Narratives and Counter-Narratives of Climate Change: North Atlantic Glaciology and Meteorology, c. 1930–1935." *Journal of Historical Geography* 35, no. 2 (2009): 237–55.

Spaar, Ilona. *Swiss Guides: Shaping Mountain Culture in Western Canada*. Vancouver: Consulate General of Switzerland, 2010.

Spence, Mark David. *Dispossessing the Wilderness: Indian Removal and the Making of the National Parks*. New York: Oxford University Press, 1999.

Stanley, Matthew. "'An Expedition to Heal the Wounds of War': The 1919 Eclipse and Eddington as Quaker Adventurer." *Isis* 94, no. 1 (2003): 57–89.

Steig, Eric J. "Another Look at An Inconvenient Truth." *GeoJournal* 70 (2007): 5–7.

Stepan, Nancy. *Picturing Tropical Nature*. London: Reaktion, 2001.

Stevenson, David, and Paula Wright. "The Living Labyrinth." *Alpinist* 67 (Autumn 2019): 52–81.

Stibal, Marek, Marie Šabacká, and Jakub Žárský. "Biological Processes on Glacier and Ice Sheet Surfaces." *Nature Geoscience* 5, no.11 (2012): 771–74.

Stuhl, Andrew. *Unfreezing the Arctic: Science, Colonialism, and the Transformation of Inuit Lands.* Chicago: University of Chicago Press, 2016.

Taillant, Jorge Daniel. *Glaciers: The Politics of Ice.* New York: Oxford University Press, 2015.

Tangborn, Wendell V., Robert M. Krimmel, and Mark F. Meier. "A Comparison of Glacier Mass Balance by Glaciological, Hydrological and Mapping Methods, South Cascade Glacier, Washington." In *Snow and Ice: Proceedings of the Moscow Symposium, August 1971,* International Association of Hydrological Sciences Publication No. 104 (1975): 185–96.

Tarr, Ralph S. *Alaskan Glacier Studies of the National Geographic Society in the Yakutat Bay, Prince William Sound and Lower Copper River Regions.* Washington, DC: National Geographic Society, 1914.

———. "The Malaspina Glacier." *Bulletin of the American Geographical Society* 39, no. 5 (1907): 273–85.

Taylor, Joseph E., III. *Pilgrims of the Vertical: Yosemite Rock Climbers and Nature at Risk.* Cambridge, MA: Harvard University Press, 2010.

Taylor, Liam S., Duncan J. Quincey, Mark W. Smith, Celia A. Baumhoer, Malcolm McMillan, and Damien T. Mansell. "Remote Sensing of the Mountain Cryosphere: Current Capabilities and Future Opportunities for Research." *Progress in Physical Geography* 20, no. 2 (2021): 1–34.

Thomas, Will. "Research Agendas in Climate Studies: The Case of West Antarctic Ice Sheet Research." *Climatic Change* 122, no. 1 (2014): 299–311.

Topham, Harold W. "The Selkirk Range, North-West America." *Proceedings of the Royal Geographical Society and Monthly Record of Geography* 13, no. 9 (1891): 554–56.

Tsing, Anna Lowenhaupt, Nils Bubandt, Elaine Gan, and Heather Swanson. *Arts of Living on a Damaged Planet: Ghosts and Monsters of the Anthropocene.* Minneapolis: University of Minnesota Press, 2017.

Tucker, Jennifer. *Nature Exposed: Photography as Eyewitness in Victorian Science.* Baltimore: Johns Hopkins University Press, 2013.

Tufte, Edward R. *Visual Explanations: Images and Quantities, Evidence and Narrative.* Cheshire, CT: Graphics Press, 1997.

Tyndall, John. *The Forms of Water in Clouds and Rivers, Ice and Glaciers.* New York: D. Appleton, 1872.

Vaux, George, Jr., and William S. Vaux Jr. "Additional Observations on Glaciers in British Columbia." *Proceedings of the Academy of Natural Sciences of Philadelphia* 51, no. 3 (1899): 501–11.

———. "Observations Made in 1900 on Glaciers in British Columbia." *Proceedings of the Academy of Natural Sciences of Philadelphia* 53, no. 1 (1901): 213–15.

———. "Some Observations on the Illecellewaet [*sic*] and Asulkan Glaciers of British Columbia." *Proceedings of the Academy of Natural Sciences of Philadelphia* 51, no. 1 (1899): 121–24.

Vaux, Henry, Jr. *Legacy in Time: Three Generations of Mountain Photography in the Canadian West.* Victoria, BC: Rocky Mountain Books, 2014.

Vaux, William S., Jr., and George Vaux Jr. "Observations Made in 1906 on Glaciers in Alberta and British Columbia." *Proceedings of the Academy of Natural Sciences of Philadelphia* 58, no. 3 (1906): 506–79.

Vetter, Jeremy. *Field Life: Science in the American West During the Railroad Era.* Pittsburgh, PA: Pittsburgh University Press, 2016.

———, ed. *Knowing Global Environments: New Historical Perspectives on the Field Sciences.* New Brunswick, NJ: Rutgers University Press, 2010.

Waddington, Edwin D. "The Life, Death and Afterlife of the Extrusion Flow Theory." *Journal of Glaciology* 56, no. 200 (2010): 973–96.

Wallace, Alfred Russel. "The Theory of Glacier Motion." *Philosophical Magazine* 42 (1871): 133–37.

Wallén, Carl Christian. "Monitoring the World's Glaciers—The Present Situation." *Geografiska Annaler* 63, no. 3–4 (2017): 197–200.

Wang, Zuoyue, and Jiuchen Zhang. "China and the International Geophysical Year." In *Globalizing Polar Science: Reconsidering the International Polar and Geophysical Years*, edited by Roger D. Launius, James Rodger Fleming, and David H. DeVorkin, 143–55. New York: Palgrave MacMillan, 2010.

Warde, Paul. "The Environment." In *Local Places, Global Processes: Histories of Environmental Change in Britain and Beyond*, edited by Peter Coates, David Moon, and Paul Warde, 32–46. Oxford: Windgather, 2016.

Warde, Paul, Libby Robin, and Sverker Sörlin. *The Environment: A History of the Idea.* Baltimore: Johns Hopkins University Press, 2018.

Warner, Michael. *Publics and Counterpublics.* New York: Zone, 2002.

Warrick, R. A., and Johannes Oerlemans. "Sea Level Rise." In *Climate Change*, edited by J. T. Houghton, G. T. Jenkins, and J. Ephraums, 257–82. Cambridge: Cambridge University Press, 1990.

Watson, Robert T., and Core Writing Team. *Climate Change 2001: A Synthesis Report.* Cambridge: Cambridge University Press, 2001.

Weart, Spencer R. *The Discovery of Global Warming.* Cambridge, MA: Harvard University Press, 2003.

Webb, Robert H., Raymond M. Turner, and Diane E. Boyer. "Introduction: A Brief History of Repeat Photography." In *Repeat Photography: Methods and Applications in the Natural Sciences*, edited by Robert H. Webb, Raymond M. Turner, and Diane E. Boyer, 3–11. Washington, DC: Island, 2010.

Weisberg, Michael. *Simulation and Similarity: Using Models to Understand the World.* Oxford: Oxford University Press, 2015.

Weller, Gunter, Matt Nolan, Gerd Wendler, Carl Benson, Keith Echelmeyer, and Norbert Untersteiner. "Fifty Years of McCall Glacier Research: From the International Geophysical Year 1957–58 to the International Polar Year 2007–08." *Arctic* 60, no. 1 (2007): 101–10.

Wheeler, Arthur O. "Affiliation with 'The Alpine Club' (London, England)." *Canadian Alpine Journal* 11 (1920): 216.

——— . "The Canadian Rockies: A Field for an Alpine Club." *Canadian Alpine Journal* 1, no. 1 (1907): 36–46.

White, Richard. *Railroaded: The Transcontinentals and the Making of Modern America.* New York: W. W. Norton, 2011.

Whyte, Kyle Powys. "Indigenous Experience, Environmental Justice and Settler Colonialism." In *Nature and Experience: Phenomenology and the Environment*, edited by Bryan E. Bannon, 157–74. London: Rowman & Littlefield, 2016.

——— . "Indigenous Science (Fiction) for the Anthropocene: Ancestral Dystopias and Fantasies of Climate Change Crisis." *Environment and Planning E: Nature and Space* 1 (n.d.): 224–42.

——— . "Is It Colonial Déjà Vu?: Indigenous Peoples and Climate Injustice." In *Humanities for the Environment: Integrating Knowledge, Forging New Constellations of Practice*, edited by Joni Adamson and Michael Davis, 103–19. Abingdon, England: Routledge, 2017.

Wilder, Kelley. *Photography and Science.* London: Reaktion, 2009.

Willen, Matthew S. "Composing Mountaineering: The Personal Narrative and the Production of Knowledge in the Alpine Club of London and the Appalachian Mountain Club, 1858–1900." PhD diss., University of Pittsburgh, 1995.

Williams, Chris. "'That Boundless Ocean of Mountains': British Alpinists and the Appeal of the Canadian Rockies, 1885–1920." *International Journal of the History of Sport* 22, no. 1 (2005): 70–87.

Williams, Raymond. *Problems in Materialism and Culture: Selected Essays.* London: Verso, 1980.

Williams, Richard S., and Oddur Sigurðsson. "Editors Introduction." In *Draft of a Physical, Geographical, and Historical Description of Icelandic Ice Mountains on the Basis of a Journey to the Most Prominent of Them in 1792–1794*, edited by Sveinn Pálsson, xviii–xxxvi. Reykjavik: Icelandic Literary Society, 2004.

Wolfe, Audra. *Competing with the Soviets: Science, Technology and the State in Cold War America.* Baltimore: Johns Hopkins University Press, 2013.

——— . *Freedoms' Laboratory: The Cold War Struggle for the Soul of Science.* Baltimore: Johns Hopkins University Press, 2020.

Wollen, Peter. "Fire and Ice." In *The Cinematic*, edited by David Campany, 118–30. Cambridge, MA: MIT Press, 2007.

Wood, Walter A. "Richard Carleton Hubley—1926–1957." *Journal of Glaciology* 3, no. 23 (1958): 277.

————. "The Parachuting of Expedition Supplies: An Experiment by the Wood Yukon Expedition of 1941." *Geographical Review* 32, no. 1 (1942): 36–55.

Worster, Donald. *A Passion for Nature: The Life of John Muir*. Oxford: Oxford University Press, 2008.

Wright, John Kirtland. *Geography in the Making: The American Geographical Society*. New York: American Geographical Society, 1925.

Yeigh, Frank. "Canada's First Alpine Club Camp." *Canadian Alpine Journal* 1, no. 1 (1907): 47–57.

Young, Neal W. "Responses of Ice Sheets to Environmental Changes." In *Proceedings of the Canberra Symposium on Sea Level, Ice and Climate Change, Canberra, December, 1979*, edited by Ian Allison, 331–60. Wallingford, UK: International Association of Hydrological Sciences, 1981.

Þórarinsson, Sigurdur. "Glaciological Knowledge in Iceland before 1800: A Historical Outline." *Jökull* 10 (1960): 1–17.

————. "Present Glacier Shrinkage and Eustatic Changes of Sea-Level." *Geografiska Annaler* 22, no. 3–4 (1940): 131–59.

INDEX

References to illustrations are *italicized*.

Appenzellar, Tim, 157–58
Arctic, Desert and Tropical Regions Information Center, 90
Arctic Institute of North America (AINA), overview, 87–88. *See also* Project Snow Cornice; Wood, Walter
Army, US, 101–2
Arrhenius, Svante, 135
Asulkan Glacier, 20

Bader, Henri, 188n74
Balog, James, 156, 158, 159
Banff, 19, 39, 178n1. *See also* Glacier House
Bates, Robert, 90
Bay in Place of the Glacier (Sít Eeti Geiyí), 60–61
Bear, Jordan, 187n57
Beckey, Fred, 191n33
belief systems, role in evidence interpretation, 139–40, 151–52. *See also* colonial perspectives; cultural context factor
Benjamin, Walter, 10, 51, 184n101
Benson, Carl, 199n6
Benson, Etienne, 178n48, 194n6
Bering, Vitus, 60
Bering Strait, 60
"The Big Thaw" (Appenzellar), 157–58
Biner, Joseph, 69
Black Rapids Glacier, 55, 126
Blue Glacier, 114, 119, 195n27
Board of Indian Commissioners, 26
Bonaparte, Roland, 47–48
Borges, Luis, 156–57, 202n31
botany and glacier studies, 187n51
boundary work, specialist language, 156
Bow Falls, 1, 3
Bow Glacier, 1–3, *4*, 164
Bow Lake Lodge, 1
Bowman, Isaiah, 74
Brady Glacier, 70
Britain, Chomolungma claims, 45
Brown, C. Suzanne, 67
Brückner cycles, 49

Bryn Mawr Glacier, 72
Buell, Oliver, 40
Burnett, D. Graham, 180
Burroughs, John, 62, 64
Burton, Pierre, 38
Bush, George H. W., 136

Callendar, Guy Stewart, 135
calving, defined, 11
Camp Century, 101–2
Camp 10 hut, Taku Glacier, 104
Canadian Alpine Journal, 46–47, 183nn88–90, 192n57
Canadian Pacific Railway: glacier study support, 31, 35, 42–43, 180n41, 182n69; government support, 38–39; mountaineering support, 31, 45, 181n60; in nationalism mythmaking, 37–39, 43, 44; tourism promotion, 37, 39–42, 182n65. *See also* Glacier House
Canadian Rocky Mountain Cordillera, area of, 19–20, *56*
Carey, Mark, 8, 166, 177n44
Carter, Jimmy, 136
Cartwright, C. F., 35, 36, 43
Cascades Range glaciers, Hubley's studies, 120–21
Century Magazine, 60–61, 185n23
CFC regulation, 200n17
Chacltaya Glacier, 158
Chakrabarty, Dipesh, 176n13, 176n24
Chambi, Martín, 171, 173
"Charney Report" (NRC), 136
Chasing Ice (Balog), 156, 158, 164
Chirikov, Aleksei, 60
Chomolungma (Mount Everest), 45, 95
Chookaneidi people (Tcukanadi people), 58–59, 185n10
Chugach Mountains, 127
Church, Phil, 117, 125
classified photograph problem, 107, 192n56
Clearwater River, 69
climate change, terminology variety, 175n3. *See also* global warming science

Fay, Charles, 42, 44
Feuz, Edouard, 35–36, 69
Field, Frederick, 68–69, 71
Field, Lila, 68
Field, William O.: on Ahlmann's work, 85; background, 68–70, 90; career span, 54–55, 67–68, 75; establishment of repeat photography program, 74; funding searches, 99, 102, 105; Glacier Bay/Lituya Bay investigations, 70–71; glacier documentation beginnings, 70–72; international studies advocacy, 92–93; invitation from Canadian Alpine Club, 183n89; Kaasteen story, 59; leadership appointment, 91; Meier's Saskatchewan research, 195n33; on Meier's sea level rise work, 144; mentoring of Post, 124; Miller's classified photograph problem, 107, 192n56; on photography's value, 109, 132; on professional identity, 147, 199n6; recognition of repeat photography's limitations, 76–77; in Sharp's reputation, 96; support for comparative work, 125; Taku Glacier study, 92–94; with tripod, 66
"50 Switzerlands in one!," 41, 181n64
Finland, 89
firn, defined, 11
firnification process, 96
fish and glaciers, 12–13
flow mechanics, 96–97, 190n29
fly-fishing, Sharp's, 96
Foote, Eunice Newton, 135
Forbes, James David, 31–32, 33–34
Forel, François, 47–48
Forest Service, US, 99
Foucault, Michel, 176n24
Fountain, Andrew, 160
Fourteenth of July Glacier, 83
Francis, Daniel, 38–39
Fraser, Don, 182n69
Fröya Glacier, 83
frozen river comparison, W. Vaux's, 35
fur traders, 60

Gagné, Karine, 151–52
Galaup, Jean-François de, 60
Galison, Peter, 178n47
Gallup polls, global warming, 151
Garrard, Rodney, 8
general circulation models (GCMs), 135–36
generalizable science, 120, 194n9
Geographical Review, 74, 84, 85
"The Geology of Mankind?" (Malm and Hornborg), 176n13
geophysical glaciology regime: overview, 15–17, 53–55, 85–86, 115; beginnings in North America, 87–89; IGY's role, 124–25; photography's role, 63, 98, 108–9, 112–13; techniques for, 108–9; as visual culture transformation, 112–14
geopolitical influences, 101–2, 137
George W. Elder, 63–64
Georgi, Johannes, 79
German-Austrian Alpine Club, 47
Gilbert, Grove Karl, 64–65, 70, 75
glacial geology theory, Muir's, 61, 62
glacial geology *vs.* glaciology, defined, 77–78
glacial theory, Agassiz's, 23, 48
Glacier Bay, 58–59, 60–62, 65, 70, 74, 185n23, 185n25
"Glacier Bay and Its Glaciers" (Reid), 70
Glacier Bay National Park, 65, 185n25
Glacier House, 19, 20, 25, 28, 30–31, 37, 42, 44, 46, 51
Glacier Ice (Post and LaChapelle), 132, 134
glacier motion: complexity of, 34, 48–50, 55, 57, 72–73, 76–77, 166; early theorizing, 33–34, 35, 48–49, 75–76; Indigenous perspectives, 58–59; Meier's perspective, 129; precision instrument based investigations, 80–84; terminology, 11; Vaux's observational approach, 35–36
glacier naturalism regime: overview, 15–16; cultural context, 16, 22–23, 25, 30–31, 34; early Alaskan studies, 62–65; European studies, 31–32; as observation, 33; repeat photograph's importance, 22–23; signif-

icance of, 51–52; tourism's role, 40–43. *See also* Vaux *entries*

Glacier Photograph Collection, 67

glaciers, overviews: Indigenous perspectives, 171–73; scientific description, 10–13. *See also specific topics,* e.g., Alhmann, Hans W.; glacier motion; Juneau Icefield *entries*

Glaciers and Glaciation (Gilbert), 64–65, 70

Glaciers of the American West, 160

The Glaciers of the Canadian Rockies and Selkirks (Vaux), 42

Glacier Studies Project, 93, 107

glacier study *vs.* glaciology, defined, 15

glaciology *vs.* glacial geology, defined, 77–78

glaciology *vs.* glacier study, defined, 15

Global Climate Change Research Initiative, 136

Global Environmental Monitoring Services, UN's, 137, 138

global warming science: CFC regulation comparison, 200n17; geopolitical influences, 137; human impact debate, 176n13; iconography of, 5–8, 23, 135, 166–67, 176n20; modeling-observation convergence, 150–51; perception problem, 151–57, 201n19; public understanding mission, 146–47, 149–50; and regional monitoring development, 137–40; research history highlights, 134–37, 196n55; scholarly approaches, 8–9, 176n13; and sea level rise research, 140–45; visual aid factor, 152, 200n16

Goldthwait, Richard, 88, 104–5

Gore, Al, 165–66

The Great Railway, 1871–1881 (Burton), 38

Green, William Spotswood, 30–31, 46

Greenland Icecap, 102, 158

Grinnell Glacier, 158

Gus George's Grocery, 99

Hall, Henry, 70

Hall, Henry S., Jr, 183n89

Hall, Marschall, 47–48

Harkin, J. B., 39

Harmon Foundation, 99

Harriman Expedition, 62, 64, 70

Harriman Fjord, 72

Harvard Glacier, 72

Harvard Mountaineering Club (HMC), 69, 70

Hayes, David, 157

Heggie, Vanessa, 103

Henderson, Alexander, 40

Henteel No' Loo' (Muldrow Glacier), 55, 126, *131*

Heusser, Calvin, 187n51

Higgs, Eric, 202n39

Hillary, Edmund, 45

historical perspective, value for examining knowledge-making, 13–15

HMC (Harvard Mountaineering Club), 69, 70

Hoeffellsjökull Glacier, 83

Hole-in-the-Wall Glacier, 93

Hoonah village, 59

Hoover, Herbert, 179n15

Hoover, Lou, 27, 179n15

Hornborg, Alf, 176n13

Howe, Joshua, 135, 153, 196n55, 201n19

Huaraz, Peru, 13

Hubley, Richard: Cascades Range glaciers, 120–21; death of, 116–17, 193n3; McCall Glacier, 194n5, 195n27; ONR report demands, 192n58; Post friendship, 117, 124, 132

Hughes, Terence, 197n74

Hulme, Mike, 152

Humble Oil (ExxonMobil), 9–10

hunting, Rocky Mountains National Park, 181n59

hydrological cycle, role of glaciers, 12

hydrology orientation, glacier research, 89, 119, 128–29, 137–38

hypothesis-evidence link, interpretation conflicts, 139–40

ice age theory, 196n51

ice cap, defined, 11

Mount Fairweather (Tsalxhaan), 70–71
Mount Hubbard, 114
Mount Logan, 71
Mount Mary Vaux, 28
Mount Saint Elias (Yas'éit'aa Shaa), 63–64, 92
Mount Saskatchewan, 68
Mount Sir Donald, 21, 32, 39, 180n34
Mount Steele, 103
Mount Vancouver, 95
Mount Victoria, 44
Muir, John, 61–62, 64
Muir Glacier, *54*, 63, 64, 71, *148*
Muldrow Glacier (Henteel No' Loo'), 55, 126, *131*
Murkami, Haruki, 203n1
Muybridge, Eadweard, 51

National Geographic, 157, 160
National Geographic Society, 63, 65
nationalism mythmaking, 37–39, 43, 44. *See also* Canadian Pacific Railway; geopolitical influences
National Research Council (NRC), 136
National Science Foundation (NSF), 125, 127, 159
National Snow and Ice Data Center, 67
Natural History of Ice (Dobrowolski), 188n74
Ned, Annie, 58, 184n6
Nepal, Chomolungma claims, 45
New Zealand, Chomolungma claims, 45
Nichols, Robert, 189n13
Nielsen, Lawrence "Larry," 122
Niitsitapiksi people, 19–20, 169
Nikon, 159
Nilsson, Einar, 90
Norgay, Tenzing, 45
Norris Glacier, 93
Norseman plane, 107, 114
Notman, William, 40
NRC (National Research Council), 136
NSF (National Science Foundation), 125, 127, 159

Nussbaumer, Samuel U., 160
Nye, John, 190n29

oblique photographs, 130
Odell, Noel, 77, 95
Office of Naval Research (ONR). *See* military support
oil spills, 132
optical unconscious idea, Benjamin's, 51, 184n101
O'Reilly, Jessica, 142
Oreskes, Naomi, 106, 203n49
Oscar Meyer Company, 99
outburst floods (*jökulhaups*), 13, 49
oxygen experiments, interpretation conflicts, 139–40
ozone layer, CFC impact, 200n17

Palache, Charles, 70
Pálsson, Sveinn, 31
Pan American World Airways, 99, 100
Parker, Elizabeth, 44–45, 46
Parker, Jean, 45
passes, railway, 31, 42, 45
Patagonia, Glacier Studies Project, 93, 105, 107, 189n13
Peirce, Charles Saunders, 187n57
Pelto, Mauri, 160
Penck, Albrecht, 32
Peru, 13, 171–73
Pfeffer, Tad, 122, 124, 158–59, 165–66
P5 rock, 32
Phillips, Norman, 136
phlogiston, combustion theory, 139
photogrammatic surveys, 182n77
Photographer's Rock, 33
photographic realism, 202n41
photography's role: in erasure of Indigenous people, 7–8, 59–60, 113–14, 169; Field's comments about, 109, 132; in geophysical glaciology regime, 63, 98, 108–9, 112–13; in Juneau Icefield Research Project, 108–12, 193n62. *See also* aerial photog-

Tarfala research program, 140, 188n69, 188n73

Tarr, Ralph, 65, 70, 129, 186n32

Taylor, Andy, 70–72

Tcukanadi people (Chookaneidi people), 58–59, 185n10

technological nationalism, 38–39

Tenth Mountain Division, 90, 92

termini fluctuations, calculation tools, 80–81. *See also* glacier motion

Third International Polar Year, 159

Thompson, Lonnie, 162

Thorarinsson, Sigurdur "Skallagrim," 141

Thorington, J. Monroe, 47

tidewater glaciers, overview, 12, 55–57

time-lapse photography, 51, 158, 159

Tlingit people (Lingít Aaní), 13, 58, 60, 61, 184n6, 185n25

tobacco regulation, doubt-mongering tactics, 153

Toboggan Glacier, 72

Topham, Harold, 32

tourism: Field's trips, 68; Glacier Bay, 61, 62; as glacier study motive, 99; railway promotion, 37, 39–40, 41–42, 182n65

Toyatte, 61

transportation barrier, cold weather research, 103

transportation networks, role in science, 57, 184n5. *See also* Canadian Pacific Railway

Tsalxhaan (Mount Fairweather), 70–71

Tsuut'ina people, 19–20, 169

Tuck, Eve, 194n9

Tufte, Edward, 161

Tulsukwe Glacier, 93

Turner Glacier, 65

Twin Glacier, 93, 104

Tyndall, John, 33–34, 135

ukukus, 171

UN Environment Programme, 137, 138

US Geological Survey (USGS), 62–63, 74, 127–28

Valdez Glacier, 72

Vancouver, George, 60

Vanderbilt, Cornelius "Commodore," 68

Van Horne, William Cornelius, 37, 40

"Vanishing Glaciers" (Miller), 106–7

Variegated Glacier, 159

varves, counting benefits, 78

Vatnajökull Glacier, 83, 96, 141

Vaux, George, Jr: as active citizen, 26; childhood, 19, 23–24, 179nn8–10; on Glacier House, 37; Illecillewaet investigations, 28–30, 33–36; photography interest, 24, 28, 179n11; photos of, *24*, *29*; railway relationship, 42–43; scientific interest, 26; success factors, 36–37; tourist-oriented visits to Rockies, 19, 28

Vaux, Mary: attitude toward Rockies, 27; childhood, 19, 23–24, 179nn8–10; civic activities, 26, 179n15; Japan trip, 179n8; marriage, 26, 179n15; mountaineering skills, 26–27; photography interest, 24, 27, 179n11; railway relationship, 42–43; tourist-oriented visits to Rockies, 19, 28

Vaux, William, Jr: attitude toward Rockies, 25; childhood, 19, 23–24, 179nn8–10; death of, 25; engineering/technical perspective, 24–25; on Glacier House, 37; Illecillewaet investigations, 35–36; photography interest, 24, 28; railway relationship, 42–43; reading lists, 43; success factors, 36–37; tourist-oriented visits to Rockies, 19, 28

Vaux family: journal contributions, 46, 48; photograph collection, 19, 20–21, 178n1; recession observation, 49–50; Wheeler relationship, 46, 183n87

Vauxite, 26

vertical photographs, 130

Vetter, Jeremy, 184n5

Victoria Glacier, 44

Virginia, 72

visual representations, value for knowledge-making, 14–15. *See also* repeat photography, overviews

WEYERHAEUSER ENVIRONMENTAL BOOKS

9 780295 752013